Challenges in Radiation Protection and Nuclear Safety Regulation of the Nuclear Legacy

NATO Science for Peace and Security Series

This Series presents the results of scientific meetings supported under the NATO Programme: Science for Peace and Security (SPS).

The NATO SPS Programme supports meetings in the following Key Priority areas: (1) Defence Against Terrorism; (2) Countering other Threats to Security and (3) NATO, Partner and Mediterranean Dialogue Country Priorities. The types of meeting supported are gene rally "Advanced Study Institutes" and "Advanced Research Workshops". The NATO SPS Series collects together the results of these meetings. The meetings are coorganized by scientists from NATO countries and scientists from NATO's "Partner" or "Mediterranean Dialogue" countries. The observations and recommendations made at the meetings, as well as the contents of the volumes in the Series, reflect those of participants and contributors only; they should not necessarily be regarded as reflecting NATO views or policy.

Advanced Study Institutes (ASI) are high-level tutorial courses intended to convey the latest developments in a subject to an advanced-level audience

Advanced Research Workshops (ARW) are expert meetings where an intense but informal exchange of views at the frontiers of a subject aims at identifying directions for future action

Following a transformation of the programme in 2006 the Series has been re-named and reorganised. Recent volumes on topics not related to security, which result from meetings supported under the programme earlier, may be found in the NATO Science Series.

The Series is published by IOS Press, Amsterdam, and Springer, Dordrecht, in conjunction with the NATO Public Diplomacy Division.

Sub-Series

A.	Chemistry and Biology	Springer
B.	Physics and Biophysics	Springer
C.	Environmental Security	Springer
D.	Information and Communication Security	IOS Press
E.	Human and Societal Dynamics	IOS Press

http://www.nato.int/science
http://www.springer.com
http://www.iospress.nl

Series C: Environmental Security

Challenges in Radiation Protection and Nuclear Safety Regulation of the Nuclear Legacy

edited by

Malgorzata K. Sneve

Norwegian Radiation Protection Authority, Østerås, Norway

and

Mikhail F. Kiselev

Federal Medical Biological Agency, Moscow, Russian Federation

Springer

Published in cooperation with NATO Public Diplomacy Division

Proceedings of the NATO Advanced Research Workshop on
Challenges in Radiation Protection and Nuclear Safety Regulation
of the Nuclear Legacy, organised jointly by NRPA and FMBA
"Ershovo" (Zvenigorod) Moscow, Russia
25-27 September 2007

Library of Congress Control Number: 2008928524

ISBN 978-1-4020-8633-5 (PB)
ISBN 978-1-4020-8632-8 (HB)
ISBN 978-1-4020-8634-2 (eBook)

Published by Springer,
P.O. Box 17, 3300 AA Dordrecht, The Netherlands.

www.springer.com

Printed on acid-free paper

PREFACE

Advanced Research Workshop on "Challenges in Radiation Protection and Nuclear Safety Regulation of the Nuclear Legacy" organised jointly by Norwegian Radiation Protection Authority (NRPA) and the Federal Medical-Biological Agency of Russia (FMBA) was held at "Ershovo" (Zvenigorod) in Moscow, Russia, 25–27 September 2007 with participants from international organisations like NATO, IAEA, ICRP and several countries Norway, Russia, United States, France, United Kingdom, Sweden and Kyrgyzstan.

The workshop was sponsored by the NATO Programme Security Through Science and Norwegian Radiation protection Authority. The sponsorship and the financial support of NATO is gratefully acknowledged.

The workshop was organized in Russia by the Federal Medical-Biological Agency of Russia and Institute of Biophysics (IBPh). The efforts of many individuals from FMBA, IBPh and other participants in producing both a technically challenging workshop are also gratefully acknowledged.

CONTENTS

SESSION II REGULATORY IMPLEMENTATION OF TREATIES, STANDARDS AND RECOMMENDATIONS

Chairmen: R. Charafoutdinov and D. Louvat

SESSION III CHALLENGES IN PRACTICAL IMPLEMENTATION OF REMEDIATION STRATEGY IN RUSSIA AND ABROAD

Chairmen: Nikolaev and D.R. Thomas

CONTRIBUTORS

Alekseeva, Valentina R.,
Regional Department No. 120, Federal Medical-Biological Agency (FMBA), 5, Biryukova street, 183060 Snezhnogorsk, Murmansk region, Russia

Barchukov, Valeriy G.,
State Research Centre-Institute of Biophysics, Zhivopisnaya ul. 46, 123182 Moscow, Russia

Deregel, Christian,
Institut de Radioprotection et de Sûreté Nucléaire (IRSN), BP 17 92262 FONTENAY AUX ROSES Cedex, France

Gordon, Boris G.,
Director of the Scientific and Engineering Center of Nuclear and Radiation Safety, (SEC NRS), 2/8 – 5, ul. Malaya Krasnoselskaya, 107140 Moscow, Russia

Gusakov-Stanyukovich, Igor V.,
Federal Atomic Energy Agency (Rosatom), Bolshaya Ordynka street 24/26, 119017 Moscow, Russia

Hawkins, Albert R.,
U.S. Department of Energy, Hanford National Laboratory, P.O. Box 99, Richland, WA 99352, USA

Kiselev, Mikhail F.,
Federal Medical Biological Agency, 30, Volokolamskoye s., 123182 Moscow, Russia

Kutkov, V.,
Russian Research Centre "Kurchatov Institute", 1, Kurchatov square, 123182 Moscow, Russian Federation, Russia

Linge, Igor I.,
Nuclear Safety Institute, IBRAE RAN, 52, Bolshaya Tulskaya street, 113191 Moscow, Russia

Mayall, Andrew,
Environment Agency of England and Wales, Team Leader (Sellafield), Nuclear Regulation Group (North), Ghyll Mount, Gillan Way, Penrith 40 Business Park, Penrith, Cumbria, CA11 98P, UK

Mazokin, Vasilij A.,
FSUE, NIKIET, p/b 788, 101000 Moscow, Russia

Mokrov, Yuri G.,
Mayak Production Association, 31, Lenina prospect, 456780 Ozersk, Chelyabinsk region, Russia

Panfilov, Alexander P.,
Federal Atomic Energy Agency (Rosatom), Bolshaya Ordynka street 24/26, 119017 Moscow, Russia

Romanov, Vladimir V.,
Deputy Head of the Federal Medical Biological Agency (FMBA), 30, Volokolamskoye s., 123182 Moscow, Russia

Salbu, Brit,
NATO and The Norwegian University of Life Sciences, Agriculture University of Norway, Isotope Laboratory, Department of Plant and Environmental Sciences, P.O. Box 5003, 1432 Ås, Norway

Shandala, Nataliya K.,
State Research Centre – Institute of Biophisics, 46, Zhivopisnaya ul., 123182 Moscow, Russia

Simakov, Anatoliy V.,
State Research Centre – Institute of Biophisics, 46, Zhivopisnaya ul., 123182 Moscow, Russia

Sneve, Malgorzata,
Senior Adviser, Norwegian Radiation Protection Authority, Department for Emergency Preparedness and Environmental Radioactivity (NRPA), P.O. Box 55, Grini Næringspark 13, 1332 Østerås, Norway

Strand, Per,
Director, Norwegian Radiation Protection Authority (NRPA), P.O. Box 55, Grini Næringspark 13, 1332 Østerås, Norway

Testov, Stanislav,
Ministry of Defence of Russian Federation, 2 Rubtsovsko-Dvorsovaja street, 119160 Moscow, Russia

Valentin, Jack,
Scientific Secretary, International Commission on Radiological Protection (ICRP), 171 16 Stockholm, Sweden

Walker, Stuart,
U.S. Environmental Protection Agency (EPA), Science and Policy Branch, Office of Superfund Remediation and Technology Innovation (OSRTI), 1200 Pennsylvania Avenue, NW (5204 P), Washington, DC, 20460, USA

Zika, Helmuth,
Swedish Radiation Protection Authority (SSI), 171 16 Stockholm, Sweden

Introduction

Introduction, Summary and Conclusions of the NATO Workshop

1 Introduction

This workshop is the second one organised by the Norwegian Radiation Protection Authority under the sponsorship of the NATO Security Through Science Programme focussing on the regulatory aspects of nuclear legacy.

The first NATO Advanced Research Workshop took place in December 2004 in Moscow on Radiation and Environmental Safety in North-West Russia and Related Use of Impact Assessment and Risk Estimation. That workshop was organised by Norwegian Radiation Protection Authority (NRPA) and Rostechnadzor. A variety of conclusions was drawn about the need for improvements in environmental risk assessment and related regulations and regulatory guidance necessary for effective and efficient supervision of nuclear legacy sites. Accordingly, a range of activities has been progressed by a number of Russian and overseas organisations which specifically address activities for remediation of SevRAO operated sites in northwest Russia. Significant among these has been the regulatory cooperation program between the NRPA and the Federal Medical Biological Agency (FMBA) of Russia.

Taking account of these developments, a second NATO workshop was held in September 2007 under the framework to consider the current challenges in radiation protection and nuclear safety regulation of the nuclear legacy. The overall objective was to share East-West competence and experience in regulatory work associated with radiation protection and nuclear safety supervision of installations built during the cold war, particularly in relation to regulatory strategies for safe decommissioning of unique or unusual nuclear facilities and remediation activities.

2 Participation

There were over 60 participants from eight countries as well as representatives from the International Commission on Radiological Protection, the International Atomic Energy Agency and NATO. The organisations involved included regulatory authorities, operators and technical support organizations. This wide level of participation reflects the importance placed upon international cooperation on nuclear legacy management issues.

3 Presentations and Papers

Session I: Nuclear Legacy Challenges
Session II: Regulatory Implementation of Treaties, Standards and Recommendations
Session III: Challenges in Practical Implementation of Remediation Strategy in Russia and Abroad
Session IV: Safety Regulation Experience in Russia and Abroad

4 Discussion and Conclusions Arising from Workshop Sessions

The following points arose from discussion of the presentations held within the four Sessions.

4.1 Session I: Nuclear Legacy Challenges

It was first noted that Norway and Russia have come together as neighbours to investigate environmental issues in an effort to find the best practical solutions to nuclear legacy challenges. The importance of regulatory support in the Norwegian Plan of Action, based on scientific evaluation of the radiological threats was emphasized. Progress has been good in developing improved regulatory documents. Now, the momentum must be maintained in the next steps. These next steps focus on independent review of safety cases and rigorous monitoring of compliance with requirements.

The new Russian Federation Unified State Programme for Handling Spent Fuel and Radioactive Waste was introduced and the steps necessary to set this up as a comprehensive programme of activities was described. It was noted that some waste management sites are in a poor state. Legacy management experience was provided from the UK, France and the USA. Information was provided on methods for site characterisation and separation of wastes into exempt, low-level and intermediate level waste, and the complication of dealing at the same time with chemical hazards, such as beryllium contaminated waste. Strategies for contaminated site management were also explained and the associated risk assessment methods outlined. Some differences in approach can be explained by site and waste specific circumstances; some others arise because legal frameworks are different. However, the reasons for differences in frameworks and related technical standards cannot always be explained, and there are other things which are different for which no obvious technical explanation. It would be interesting to investigate the reasons (cultural, economic, political et cetera, and to ask if there are any material affects on levels of safety and human and environmental health protection. The development of locally relevant standards based on international guidance, rather than universal application of detailed international prescriptions, is noted by the Nuclear Energy Agency.[1] It could be useful to investigate how far the details can be derogated to regional and local responsibility without significantly compromising safety principles, e.g. as recently consolidated by the International Atomic Energy Agency.[2]

4.2 Session II: Regulatory Implementation of Treaties, Standards and Recommendations

Session II started by a scene setting of the present international regime governing radiation protection and its near and longer-term evolution. It was noted that Russia has started to be actively involved in international nuclear safety instruments and has introduced into its present set of regulations the main elements of the ICRP radiation protection system. It was also noted that the present Russian regulatory system for nuclear safety and radiation protection has not yet been harmonized with international standards and not yet been independently reviewed. Some issues like the multiplicity of regulatory bodies and of regulatory functions or the application of the principle of optimization in radiation protection would deserve more attention in the near future.

In the succession of presentations made to illustrate regulatory processes in different countries developing large nuclear programmes, critical differences were evidenced: performance based approach *versus* prescriptive approach; dilution of regulatory functions *versus* integrated single regulatory body; risk informed and documented licensing process *versus* "*ad hoc*" agreement based on technical grounds. Experience in different countries in achieving an appropriate balance could be used as lessons learned to further improve and strengthen the Russian regulatory system.

[1] 'Radiation Protection in Today's World: Towards Sustainability.' Nuclear Energy Agency, Organisation for Economic Cooperation and Development. Paris.

[2] 'Fundamentla Safety Principles' International Atomic Energy Agency. IAEA Safety Standards for Protecting People and the Environment. Safety Fundamentals No. SF-1. Vienna.

4.3 Session III: Challenges in Practical Implementation of Remediation Strategy in Russia and Abroad

Presentations included challenges in remediation of sites in Russia and the experiences of regulators and operators in the Unites States and France.

• *Remediation Challenges at SevRAO Facilities*

SevRAO comprises three branches in NW Russia: Andreeva Bay, Saida Bay and Gremikha. Priorities facing SevRAO are the remediation of Andreeva Bay and Gremikha Sites of Temporary Storage, including preparing the right conditions for removal of SNF and management of radioactive waste. The international community is helping SevRAO overcome these challenges with support from Norway, United Kingdom, Sweden, Italy and the EBRD for Andreeva Bay. France is providing support at Gremikha and Germany had funded the construction of a concrete pad to store some 100 defueled submarine reactor cores 70 years or more for each unit. Further waste management activities are planned for Saida Bay.

Some of the main issues discussed and seeking solutions include:

- Removing undamaged SNF
- Shipment and safe storage of SNF to Mayak for processing
- Damaged SNF to be removed with funding from EBRD
- Storage of SRW and LRW
- Improving health and safety of personnel working at the sites
- Improving methodologies on materials accounting
- Examination of Type 6 containers
- Proper identification of materials, before shipment
- Implementation of measures to prevent unauthorised use of SNF
- Rehabilitation of sites to brown fields status

More broadly, it is recognised that the development of operations and facilities at Saida Bay needs to be linked to an overall radioactive waste management strategy for Russia.

• *Challenges in Radiation Protection and Nuclear Safety Regulation at Hanford*

The US Department of Energy's perspective was provided on how they handled their nuclear legacy at Hanford through experiences and lessons learnt about compliance with the current Regulatory Environment and how waste is stored at the Restoration Disposal Facility with its multi-layer liner system. The penalties for non-compliance can be severe. This presentation reminded operators that they are obliged to comply with the Regulators. However, while the use of sanctions against bad practice is an appropriate mechanism in some circumstances, it was suggested that the regulator needs to encourage openness from operators and to reward good behavior in response to difficult situations, so that errors do not remain hidden.

• *US EPA Superfund Radiation Policies*

The United States Environmental Protection Agency addresses site cleanup through its:

- National Contingency Plan
- National Priorities List
- Comprehensive Environmental Response, Compensation & Liability Act, CERCLA or "Superfund"

Overviews were provided on the CERCLA process and the different Superfund sites, key CERCLA guidance and tools that specifically address radionuclides, and EPA's relationship to Department of Energy and NRC site cleanups. This was another interesting approach to the regulatory process.

• *Site Experience from Centre de Cadarache*

The French experience showed how they had recovered historical waste stored in five trenches between 1969 and 1974 (within IAEA guidelines) as an experimental temporary storage facility for low level radioactive waste. All waste stored in the trenches had to be retrieved, processed and packaged for final disposal in licensed disposal facilities managed by ANDRA. Low level short lived waste was stored at Centre de Stockage de l'Aube and very low level waste was stored at the Centre de Stockage TFA de Morvilliers. Also described were the facilities in operation in France for the incineration, melting and recycling of radioactive waste.

• *Experience of DalRAO in Remediation Problem Solving*

DalRAO (Vladivostok) has similar issues to SevRAO (NW Russia) in the remediation of its site. An analysis on their research on the levels of contamination and rehabilitation of buildings and sites conducted in 2001 through to 2007 included approaches on liquid radioactive waste storage, retiring contaminated reservoirs, determination of the radiation environment in and around facilities, and storage of defuelled submarine reactor compartments. Provision of support from Japan and Germany was noted.

• *Study of Challenges in Protection of Radiation Hazardous Facilities at SevRAO Sites*

This presentation highlighted the existing Russian legislation on waste management at hazardous radiation facilities in Russia compared with the IAEA classification on RW. It also covered quantitative guidance on: exempted waste, very low level radioactive waste (VLLW), categories of industrial wastes being generated, criteria for industrial waste categorization according to the combination of radiation and chemical factors, algorithms of VLLW management, waste composition and the nature of contamination of SNF and radioactive waste in Andreeva Bay, concluding with criteria for site release from regulatory control.

It is notable that previously, the Russian system did not have standards in place for control of these decommissioning and remediation activities.

- ***Strategic Planning of Environmental Remediation of Temporary Storage of SNF and RW in Gremikha Village***

This presentation was about developing a concept and strategy of ecological rehabilitation at Gremikha. This included options on storage of SNF and radioactive waste, and/or a repository for disposal of radioactive waste. It also included discussion of the brown field concept as applied to a radiation industrial facility or non-radiation industrial facility taking into account existing Russian legislation, availability of required technical support and experience as well as the health and safety of the public and site workers over the period of time of remediation.

It was noted that such complex problems arise at all major nuclear legacy sites.

4.4 Session IV: Safety Regulation Experience in Russia and Abroad

The FMBA presented the main regulatory challenges in the Russian legal framework. Some components have existed only a few years while others, like the Institute of Biophysics (IBPh), have been long established. Numerous regulatory documents had been issued over the last few years, by parliament and government concerning the responsibilities of FMBA, but also by FMBA, on the responsibilities of licensees. As expected in a large federation, there were some tensions between agencies and between administrative levels, because what is good practice with respect to one protection objective may be not so good for another. In response to a question, it was confirmed that the recruitment of qualified personnel was difficult at times.

FMBA provided a more specific description of challenges in radiation protection regulation at SevRAO facilities. There were regulatory problems at several sites in the region, and the regulatory responsibilities span a spectrum, from the management of spent fuel from nuclear-powered ice-breakers and other crafts to rehabilitation of contaminated areas. Many of the problems had occurred because storage buildings for radioactive materials had been used well beyond their useful service life, as evidenced by evocative photos of dilapidated buildings.

IBPh then discussed the radiological protection of workers in operations at SevRAO facilities. It was said that no problems in occupational radiological protection were due to design flaws. Instead, problems occurred because of accidents (not necessarily ‘radiological’) and other disruptions of normal procedures. Efforts to return sites to a brown field condition were hampered by lack of information about radiation and other physical conditions, possible fuel leakage, etc., as well as by the past use of unique and ‘irregular’ hardware and instrumentation. Innovative methods had been used to provide solutions to these retrospective problems.

Another presentation from IBPh gave a fascinating account of radiological protection of the public and environment near the location of SevRAO facilities. Based on the usual principles of justification, optimisation, and application of dose limits, constraints and reference levels had been established to ensure acceptable levels of exposure. Current actual exposures of the public in the Murmansk area due to radiation facilities were low, in the order of 0.03 mSv per year, but that there were large potential risks because of the amounts of highly active waste. After an initial threat assessment and analysis of the situation, reference levels representing the maximum that the regulator would plan to allow for had been set, and it was stressed that, in line with the new ICRP Recommendations,[3] optimisation was expected to lead to actual clean-up levels below the reference values.

A further presentation from IBPh described medical and radiological aspects of emergency preparedness and response at SevRAO facilities. This outlined a framework for emergency preparedness planning at federal and local levels, focusing primarily on medical handling of various kinds of casualties. The results of inspections that had been performed on site were that documents covering all the important topics were in place, but that the clarity and level of detail in these documents were not sufficient. There was training and exercises and as ever in such contexts, 'more of the same' was required. It was concluded that an adequate infrastructure had been created, but several aspects needed to be enhanced. In response to a question, it was confirmed that there were many other aspects of emergency preparedness than the medical ones presented, and that these also were planned and exercised.

A final IBPh presentation described current norms and standards for supervision of waste management at SevRAO facilities. This showed that Russia has the usual hierarchy of document levels: laws by parliament, radiation protection regulations by government, radiation protection requirements and guides by agencies and supporting organisations. In collaboration with IAEA, waste classification was being developed, taking 'general' waste handling principles into account. Some of the industrial waste concerned is also chemotoxic. Echoing some of the earlier presentations, the difficulties posed by past irregularities were mentioned, such as undefined mixing of various classes of waste and infrastructures that had not been properly maintained. A comment stated that while the system and numerical values appeared reasonable, there was no overall safety assessment and that this, taking local site and waste specific factors into account, would be crucial.

The Swedish Radiation Protection Authority gave a brief presentation of the Swedish regulatory system for handling radioactive waste, and showed an interesting video outlining how the operator plans to dispose of spent fuel and other HLW. He mentioned that significant funds were available for the purpose; it was commented that these had been accumulated from rate-payers because of the foresighted decision to levy an extra charge for future waste handling from the introduction of nuclear power in Sweden.

[3] 'The 2007 Recomendations of the International Commission on Radiological Protection.' ICRP Publication 103. Annals of ICRP, volume 37, Nos. 2–4, 2007. Elsevier.

5 Overall Conclusions

From the presentations and discussions it can be concluded that the Russian standards are generally consistent with international recommendations, but that there is scope for continuing improvement of regulatory processes and procedures, as well as the need for more appropriate norms and standards to manage special situations. In part these needs and observations arise because of the unusual conditions at SevRAO sites, but they also arise at other Russian sites and in similar sites in other countries, so that continuing cooperation can be useful in a wider Russian context and in other countries. It may also help international agencies to develop more practically effective recommendations and guidance.

Particular problems arise in the decommissioning of uranium mining and milling facilities, Long term policies for land use, contaminated land management and hazardous and radioactive waste management present easily expressed multiple objectives but they present complex risk management challenges. For example, policy suggests that further legacies should not be created for future generations to manage, but early action may create additional hazards now.

It can also be concluded that the reason for poor conditions at some sites has been the lack of, or poor development of, a broad safety culture involving all workers at all levels in safety management. Interactions at the technical level between relevant organisations do take place, but the structures under which they occur are not very flexible, and the processes for local, regional and federal coordination could be made more effective.

While development work suggested above is clearly to be valued, at the same time, those with the specific responsibility must be ready to provide vigorous supervision of current and planned operational projects in a timely and effective manner.

Regulatory processes must be clear and readily interpreted, so that all partners know what is required. This can be achieved by early prescription of requirements. At the same time, the inherent inflexibility in such an approach can lead to difficulties in managing new information, whether this is about the wastes themselves, the local environment or changes in safety and protection objectives. A suitable balance has to be sought.

While the regulator has to be able to take firm action by the use of sanctions and the courts to correct errors and omissions in on the part of operators, there is also a need to promote and encourage operators to come forward with recognition of possible past failures. Good behaviour should be rewarded.

An important question was raised concerning how to organize interfaces among interest groups concerned with legacy management:

- Scientific and technical evaluation – different scientific and engineering disciplines
- Regulatory approach – inter-regulatory cooperation on safety, human health and environmental protection
- Practical solution – operators, waste producers and waste managers
- Political situation – politicians, local and regional representation
- Public acceptance – local and regional public interest groups

There is a need for an efficient process for:

- Working together
- Managing information flow
- Getting the balance right in multi-attribute problem assessment, while
- Not forgetting the separate responsibilities of each interest group
- And not using complexity as an excuse to do nothing

A proposed starting point, suggested by the Environment Agency of England and Wales (EA), is to adopt an agreed set of principles. The intention is that all involved have shared objectives and the early dialogue reduces the chance of having to make corrections and changes later. Furthermore, *to achieve* the best environmental results, the EA traditional regulatory activity such as licensing, compliance assessment and enforcement, has needed to be supplemented with partnership working and regulatory advice and influence.

6 Recommendations

The Russian Federation has responsibility to manage its own nuclear legacy. But it is also one of several countries in the global network of nuclear activities. Harmonisation of approaches is valuable in building future cooperation, but local conditions may influence the best local solution. Accordingly, future exchanges, such as those provided for by this workshop, should be encouraged.

Development of a broader and deeper safety culture should be a long term objective, while at the same time maintaining the highest standards of radiation protection and nuclear safety as possible.

There are many complex issues to be addressed and they cannot all be solved at once. Clear recognition of the major threats, as well as weakness in regulatory processes, can be useful in directing future resources. However, at this stage it is clear that there are specific regulatory issues to address with respect to regulatory requirements and guidance for nuclear legacy sites concerning:

- Site remediation
- Waste forms for long term storage and disposal and
- Disposal facilities

In turn, such work is dependent on better radioactive waste and contamination characterisation, as well as site characterization information.

Such guidance needs to be thoroughly based on the best use of scientific and technical information. At the same time, part of the solution relates to policy issues and value judgements, and so broader interaction among regulators, operators and other stakeholders is to be encouraged.

International Cooperation of FMBA RF Aimed at Radiation Safety Assurance in Northwest Russia When Solving Nuclear Legacy Challenges

M.F. Kiselev

Nuclear power engineering development is an important success of the mankind. Nevertheless, despite its great benefit, management of the ionizing radiation sources results in a number of problems. One of the relevant challenges is connected with assurance of human and environmental protection against ionizing radiation and radionuclides resulted from nuclear reactions.

Now, in the light of intensive development of nuclear power engineering and taking account of increasing amount of States possessing technologies of the fissionable atom management, the radiation protection problem is more relevant. The significance of this problem increases when we talk about the future of the mankind and about health preservation of our offspring.

The international community recognizes its responsibility in this issue in full, and takes actions for organization of cooperation between the States and international organizations directed to safety improvement of the spent nuclear fuel (SNF) and radioactive waste (RW) management. The projects being implemented within the "Agreement on Multilateral Nuclear Environmental Program in the Russian Federation" serve as a positive example of such cooperation. This Agreement became a logical continuation of the initiative accepted by the Scandinavian countries and Russia under the name the "Northern Dimension Environmental Partnership".

Today, the most significant projects implemented within the "Agreement on Multilateral Nuclear Environmental Program in the Russian Federation", are as follows:

- The infrastructure restoration at the former shore technical base in Andreeva bay
- Increasing of productive potential of the transport technological system for discharge and management of the SNF
- Decommissioning of multi-target submarines
- Development of the innovation technologies for the SNF temporary storage and solid waste treatment
- Generation of radioecological monitoring hardware
- Arrangement of the terrestrial facility for the reactor compartment store in Sajda bay

Federal Medical-Biological Agency (FMBA), Russia

M.K. Sneve, M.F. Kiselev (eds.) *Challenges in Radiation Protection and Nuclear Safety Regulation of the Nuclear Legacy*,

Extreme radiation conditions existing at SevRAO enterprise in Murmansk region need irregular approaches in the course of supervision and regulation of this facility operation.

Having in mind the "Agreement on Multilateral Nuclear Environmental Program in the Russian Federation", The Government of the Kingdom of Norway developed the Plan of Actions for work implementation in Northwest Russia. This Plan includes not only operations aimed at decommissioning of nuclear submarines, but also efforts directed to improvement of regulation at the SNF and RW management.

The Federal Medical-Biological Agency (FMBA RF) is an authorized regulatory body of the Russian Federation dealing with supervision of radiation safety of the personnel working at Rosatom enterprises and the people living in the territories falling under hazardous effect of above facilities. One of the most important aspects of FMBA RF regulatory responsibilities relates to remediation of the sites for SNF and RW storage at SevRAO facilities in Andreeva bay and Gremikha village on the Kola Peninsula.

As elaboration of the Plan of Actions in the Northwest Russia, developed by the Norwegian Government, in October 2002, the "Protocol of cooperation between NRPA and Federal department "Medbioextrem" (now – FMBA RF) was signed. The main goal of this cooperation is an implementation of the projects aimed at FMBA RF performance of effective and efficient regulatory sanitary hygienic supervision of radiation safety and protection assurance during operation at SevRAO facilities. Within this cooperation, six projects had been accomplished and other four projects are under implementation now, dealing with different aspects of supervision and control.

The overall goal of the projects is to evaluate radiation risks at SevRAO facility and to develop some regulative documents regarding effective regulatory supervision procedure with respect to radiation protection of workers and the public as well as environmental protection during operations in routine, abnormal and emergency situations.

In order to inform the community about the mutual efforts in the field of regulatory supervision, the popular brochure had been prepared containing the detailed presentation of the findings of investigations.

Within the projects completed, the Guidance had been issued "Hygienic requirements for radiation protection of workers and the public during planning and arrangement of SNF and RW management at SevRAO facility No 1" and the Guidance «Criteria and norms of remediation of SevRAO (of the Federal Atomic Energy Agency) sites and facilities contaminated with man-made radionuclides».

At the present time, NRPA-FMBA cooperation is successfully continuing. With the purpose to improve regulatory functions of FMBA RF supervision bodies, two guidance documents are now under development, in particular: «Radiological and medical criteria to initiate urgent protective measures» and «Hygienic requirements for radiation safety assurance during industrial waste management at SevRAO facility», as well as four guidelines «Radioecological monitoring on-site STS and in the supervised area in the course of SevRAO STS conversion», «Requirements for the personal dose monitoring of the occupational exposure at SevRAO facility No 1», «Regulation of radiation monitoring at SevRAO facility No 1» and «Special features of ALARA principle application during SNF and RW management at SevRAO facility No1».

In addition to the regulative activity within bilateral NRPA-FMBA cooperation, work aimed at exchange of experience in the field of regulatory supervision is performed under active support of NATO Secretariat for public diplomacy. The results of such kind of activity are connected with traditional holding of international workshops devoted to discussion of different aspects of radiation safety and protection. This workshop serves as an example.

Moreover, by NRPA request and with approval of the Consultative Group for chemistry/biology/physics, the NATO Assistant Secretary General for public diplomacy assigned a Grant in support of the Project "Exchange of experience in the sphere of regulation to reduce the risks associated with operation of nuclear facilities". Within the framework of the Grant, experts from the FMBA of Russia, State Research Center "Institute of Biophysics" and "South-Ural Institute of Biophysics" visited the United Kingdom and the USA.

The purpose of the working visits were to exchange experience and to familiarize the experts with the structure of the radiation safety regulatory organizations in the UK and USA, their working principles and organizational methods, nuclear industry operators of these countries within a period of decommissioning and the system of RW management and remediation of radioactively contaminated territories.

The experience gained during these visits finds its reflection in above-mentioned regulative documents.

Summarizing the above reviewed, we can discover obvious positive findings of cooperation between NRPA of the Kingdom of Norway and FMBA of Russia in solution of challenges in radiation safety and protection assurance in Northwest Russia. The regulative projects accomplished in the reviewed period permit not only to ensure implementation of effective state sanitary epidemiological supervision of radiation hazardous operations at SevRAO facilities, but also to promote coordination of works in the field of radiation protection and safety.

Complicated problems of radiation protection assurance for workers and the public will be considered in the proceedings of the workshop. I believe that the workshop will be successful, because many leading Russian and foreign scientists, operators and regulators participate in it.

Welcome of Federal Atomic Energy Agency

A.P. Panfilov

Dear participants of the Workshop!

In the name of the Federal Atomic Energy Agency of the Russian Federation, let me welcome you to Eshovo.

The important challenges in radiation protection and nuclear safety regulation of the nuclear legacy are to be discussed within this International Workshop, which begins its work today. These challenges are very relevant for the Russian Federation given the need to solve problems resulting from the past nuclear activities and the current renaissance of the nuclear power in our country.

In October of 2006, the Government of Russian Federation (RF) issued a directive aimed at intensive development of atomic powered engineering and industry of Russia in the next few years and for the period till 2015. We plan significant financing of the Program – 1,471.4 milliard rubles,[1] including 674.8 milliard rubles out of proceeds of the federal budget, which will be intended mainly for construction of NPP power units.

The RF Government has adopted the Federal Target Program of nuclear and radiation safety assurance for the period to 2015, which has became an important step in managing the nuclear legacy. The amount of this program financing is 145.3 milliard rubles, including 131.8 milliard rubles out of proceeds of the federal budget. The representative of Moscow Nuclear Safety Institute, IBRAE RAN, Mr. I. Linge, will make a special presentation within the Workshop devoted to this program.

Having in mind the fulfillment of international engagements of the Russian Federation concerning reduction of armaments, special attention is paid to works aimed at dismantlement of nuclear submarines and their serviced ships being decommissed from the Navy.

According to the RF Government directive, in 1998, two special enterprises were arranged within the RF Minatom structure – FSUE "SevRAO" and FSUE "DalRAO", – and responsibility for four former shore technical bases of the Navy has been transferred to these enterprises. The majority of buildings and sites at these shore bases need decontamination and remediation. In addition, the Spent Nuclear Fuel (SNF)

Federal Atomic Energy Agency (Rosatom), Russia

[1] US$1 is about 24.4 rubles.

M.K. Sneve, M.F. Kiselev (eds.) *Challenges in Radiation Protection and Nuclear Safety Regulation of the Nuclear Legacy*,

and Radioactive Waste (RW) storage facilities of SevRAO and DalRAO, operations have to be improved under difficult conditions, which do not meet the current safety requirements. Successful implementation of remediation work under abnormal conditions requires a set of measures to be introduced to assure occupational protection and public safety, including generation of the special regulatory documents.

International cooperation and assistance of foreign States play a significant role in this issue, including exchange of experience both in decommissioning of nuclear military and industrial facilities, and RW management. Improvement of regulatory procedures is very important for enhancement of nuclear and radiation protection and safety. Much is already implemented in northwest Russia, where active collaboration is developing between Russia and Norway, UK, USA and other States.

This Workshop is another significant measure aimed at experience exchange in radiation protection and legacy management. Obviously, the participation of representatives from different international organizations, and leading researchers and practitioners from different States will allow making success more significant in the field of occupational and public radiation protection assurance.

I wish all participants of the workshop to be hard working and successful in their business and to enjoy a pleasant stay in such a lovely place near Moscow!

NATO Support to Non-military, Civil Science for Peace and Security

B. Salbu

Abstract The civil NATO has supported non-military, civil science since the 1950s. With time, NATO has been enlarged with a number of new states, and the civil part of the organisation has developed and has been reorganized. The last reorganization took place in 2006, when the Science for Peace and Security (SPS) Committee was formed. The present paper describes the new organization and the priorities of relevance for radiological and nuclear research topics.

Keywords Civil science, SPS, defence against terrorism, scientific collaboration, Environmental Security Panel (ESP), TENORM

1 NATO Support to Science

The support to non-military, civil science from NATO was initiated in the 1950s based on initiatives from Canada, Norway and Italy, to allow scientist to have a dialogue across the "iron curtain" during the cold war. In 1992, at the end of the cold war, NATO was enlarged with some new member states. Furthermore, collaboration was established between NATO and the Russian Federation, forming the NATO-Russia Council (NRC). Since then, the enlargement of NATO has continued and following the Inaugural Meeting 20 October 2006 NATO represents at present 26 countries (Fig. 1).

In NATO, the organization of the civil science is an integral part of the Public Diplomacy Division (PDD). The objectives of PDD include public information in general and the civil science cooperation such as the Science for Peace and Security (SPS) as well as the Reaching out to civil societies. This includes the collaboration between NATO countries, Partner countries, as well as the Mediterranean Dialogue (MD) countries.

Isotope Laboratory, Department of Plant and Environmental Sciences, Norwegian University of Life Sciences, Norway

M.K. Sneve, M.F. Kiselev (eds.) *Challenges in Radiation Protection and Nuclear Safety Regulation of the Nuclear Legacy*,

Fig. 1 The Inaugural Meeting 20 October, 2006 a the NATO Headquarters, Brussels (Photo: NATO)

In 2002, NATO was also associated with the Environment and Security Initiative in collaboration with international organizations such as ENVSEC, UNDP, UNEP and OSCE. The aim of the initiative is to initiate and co-ordinate projects in environmental security, mainly in Central Asia and the Caucasus region.

2 SPS Committee

The Science for Peace and Security (SPS) Committee was established June 28, 2006. The SPS was established as the primary NATO body over a programme for enhancing cooperation with all partners based on science and innovation and should conduct activities aligned with NATO's strategic objectives and especially the priorities of the partners. The SPS has a 'horizon-scanning' role in identifying future threats, raising awareness and finding solutions. The SPS includes four advisory scientific panels, including the Environmental Security Panel (ESP). Furthermore, expert groups will be established within high priority areas, The Nuclear/radiological Expert Group (NREG) was established in 2007.

The SPS programme is based on non-military, civil science cooperation and key priorities of the involved countries where the programme elements are either nationally Funded Activities (former CCMS) or NATO Managed Activities. The SPS objectives include:

- Establish concrete civil collaboration between NATO countries and Partner and MD countries
- Contribute to solving problems effecting large societies in Partner and MD countries
- Contribute to stability and peace e.g. by promoting regional co-operation
- Promote NATO's values and image in targeted communities in Partner and MD countries and society at large
- Young generation of 'Leaders of Tomorrow'
- Provide seed money for seed projects to provide the basis for addressing priority needs

A unique network of cooperation has been established between NATO countries, Partner countries and the Mediterranean Dialogue (MD) countries. The 26 NATO Countries include Belgium, Bulgaria, Canada, Czech Republic, Denmark, Estonia, France, Germany, Greece, Hungary, Iceland, Italy, Latvia, Lithuania, Luxembourg, Netherlands, Norway, Poland, Portugal, Romania, Slovakia, Slovenia, Spain, Turkey, United Kingdom and United States. The 23 Partner Countries include Albania, Armenia, Austria, Azerbaijan, Belarus, Bosnia & Herzegovina, Croatia, Finland, Georgia, Moldova, Montenegro, Ireland, Kyrgyz Republic, Kazakhstan, Russia, Serbia, Sweden, Switzerland, Tajikistan, the former Yugoslav Republic of Macedonia, Turkmenistan, Ukraine and Uzbekistan. The seven Mediterranean Dialogue Countries include Algeria, Egypt, Israel, Jordan, Mauritania, Morocco and Tunisia. In total 56 countries are collaborating within the SPS programme.
The SPS key priorities are:

1. Defence Against Terrorism:
    - Rapid detection of CBRN agents and weapons, and rapid diagnosis of their effects on people
    - Novel and rapid methods of detection
    - Physical Protection against CBRN agents
    - Decontamination of CBRN agents
    - Destruction of CBRN agents and weapons (e.g. chemical & vaccine technologies)
    - Medical countermeasures
    - Explosive detection
    - Food security
    - Information security
    - Eco-terrorism countermeasures
    - Computer terrorism countermeasures
2. Scientific Collaboration to Counter Other Threats to Security:
    - Environmental security (e.g. desertification, land erosion, pollution)
    - Water resources management
    - Management of non-renewable resources

- Modelling sustainable consumption (e.g. food, energy, materials, fiscal measures and environmental costing)
- Disaster forecast and prevention
- Human and societal dynamics (e.g. new challenges for global security, economic impact of terrorist actions, risk studies, topics in science policy)

3. Partner Country Priorities:

   - Specific topics for collaborative research have been identified by the Partner countries, High priorities for individual Partner countries.

Some themes such as Environmental Security, Computer Networking, Counter-Terrorism are common for all areas.

3 Environmental Security Panel (ESP)

Key areas of relevance for ESP are Natural hazards, Human induced environmental hazards, and Degradation of the World's Natural Resource Base, as outlined by the panel in 2006. As stated by the panel; the environmental security is being threatened by policies and practices that affect human health and which reduce or degrade the world's natural resource base. Further, while population growth and urban poverty are factors that affect environmental security, the immediate causes are human-induced pollution and poor management of natural resources.

Natural Hazards: Science and technology should support disaster reduction capabilities to enable societies to be more resilient to natural hazards, and ensure that the development efforts do not increase the vulnerability to these hazards.

Human Induced Environmental Hazards: Humans are increasingly affecting the natural environment through impacts such as acid rain, fallout from a nuclear or other trans-boundary contamination (e.g., Chernobyl accident), chemical pollution (e.g., heavy metals and POPs), inadequate water management policies (e.g., the virtual disappearance of the Aral Sea), health-endangering air pollution (e.g., fires in over-extended forest management practices in Southeast Asia) and massive destruction of arable land through unsustainable agricultural practices.

Degradation of the World's Natural Resource Base: Many existing practices and policies where Science and technology play a role have the unintended consequence of degrading the world's natural resource base, such as: deforestation and land degradation, improper land use and ineffective management of rivers and coastal areas, changes in temperature and sea-level rise from climate change and variability, unplanned urbanization, human vulnerabilities and increasing impoverishment in developing countries, and increasing urban and rural infrastructure vulnerabilities from natural hazards and globalization changes, including transportation, water, electric, gas, drainage, storage facilities and communication networks.

Thus all aspects related to these hazards; hazard identification, impact and risk assessments as well as countermeasures, will be relevant topics for the ESP.

4 Target Participants and Mechanisms

The technological programmes and activities of NATO are based on a "bottom-up" strategy where proposals are invited from the scientific and technological communities in order to address relevant scientific questions, both cutting edge questions that evolve from the science itself and opportunities that promote collaboration among scientists and technological experts from all NATO countries and those affiliated with it. NATO For problem-driven and solution-oriented issues related to the highest identified priorities of NATO, "top-down" initiatives that can promote and support through the science and technology programmes can also be established. These programmes should provide effective solutions to urgent environmental and natural resource-oriented problems, particularly in countries or regions where security and stability are at risk.
The programme includes several mechanisms to achieve the goals:

- For small research groups: Collaborative Linkage Grants (CLGs) to work together for security
- For large groups of experts: Workshops (ATC-ARW) on security-related priorities or educational training (ASI)
- For Institutions and established research groups: Science for Peace Projects to conduct joint security R&D and upgrade Partner laboratories

For Societies at large, projects with NATO support can be initiated to solve problems affecting societies as a Nationally Funded Activities (NFA, Former CCMS). The NFA proposals can be presented by any NATO, Partner or Mediterranean Dialogue countries prepared by national authorities and submitted to the SPS Committee by the national SPS member. For national or regional priorities of interest to several nations, Pilot Studies (3–5 years), Short-term projects (Maximum 18 months), and Ad-hoc workshops can be initiated.

The dissemination of information from the SPS activities is essential to NATO. Results from Advanced Research Workshops (ARWs) and Advanced Study Institutes (ASIs) are published in books under the NATO Science for Peace and Security Series. A newsletter is published quarterly, on-line publications are presented on the Programme website: www.nato.int/science, and Partnership Real-Time Information is provided via the Management and Exchange System: ePrime.

5 Relevance to Radiation Protection and Nuclear Safety Regulation of the Nuclear Legacy

There is a significant number of nuclear and radiological sources in NATO countries, Partner countries, as well as the Mediterranean Dialogue (MD) countries, which have contributed, are still contributing, or have the potential to contribute to radioactive contamination in the future. Key sources and contaminated sites of concern are nuclear weapons tests including safety trials and peaceful nuclear explosions preformed for instance in central Asia. Technologically enhanced levels of naturally occurring radionuclides (TENORM) due to U mining, tailings and production are also of concern. As a legacy of the cold war and the nuclear weapon programme in the former USSR, thousands of square kilometers in the Central Asia countries are contaminated. Furthermore, large amounts of scale from the oil and gas industries contain sufficient amounts of TENORM. A series of nuclear reactors, to be decommissioned or still in operation, represents also nuclear risks as well as poorly controlled storage of spent nuclear fuel and other radioactive wastes.

In the assessment of nuclear risks, the probabilities of accidents and their consequences are assessed. The analysis includes evaluation of the sources and possible accidental scenarios, ecosystem transfer, biological effects as well as social and economic consequences following the event. Sources may occur stationary (point sources), temporally (labile sources), and outside an individual country. In general, the larger the inventory of radionuclides the greater the hazard, unless specific safety precautions are taken. In most cases, the sources are known and inventories (Becquerel, Bq) are well established, while in other cases (e.g. old waste storage facilities) the information may be less complete or lacking. When it comes to unforeseen events such as sabotage and terrorisms, neither the source (inventory) nor the localities are known. However, risk assessments and priorities of key sources of concern can be utilized to introduce more safety measures and to build up a relevant emergency response. Thus, environmental impact and risk assessments can form the basis for practical policy-making, such as authorization of industrial releases, interventions within highly contaminated areas, countermeasures (e.g., food restriction), clean-up strategy and remediation of contaminated areas, as well as the updating of legislations and laws associated with the radiation protection of man and the environment. As several countries are facing similar nuclear risks and because the contamination is transboundary, regional and international co-operations within this field seems highly relevant. Thus, nuclear and radiological risks are well within the scope of NATO SPS.

Several relevant Advanced Research Workshops (ARWs) have been organized recent years, for instance the ARW on Nuclear Risks in Central Asia, was organized in Almaty, June, 2006, with many participants from the Central Asia region. Furthermore, the ARW on Hot Particles Released from Different Nuclear Sources was organized in Yalta in May 2007. NATO has also recently demonstrated that nuclear issues are a key priority as the NATO Science Partnership Prize for 2007 was awarded to Professor Nick Priest (UK) and Professor Mukash Burkitbayev (KAZ) for the NATO Science for Peace Projects for the Semipalatinsk Project SEMIRAD.

6 Conclusions

The SPS, ESP and the expert group should be effective tools for initiating transboundary, regional and international co-operations between NATO, Partner or Mediterranean Dialogue countries within environmental security issues. Problems associated with nuclear and radiological risks are well within the scope of NATO SPS priorities. As several countries are facing similar nuclear risks and because the contamination is transboundary, regional and international, bottom-up or top down initiatives within this area should be encouraged.

Acknowledgement Dr. Deniz Yüksel-Beten, Head, Threats & Challenges Section, Public Diplomacy Division, NATO (www.nato.int/science).

Introduction – Norwegian Perspective on Nuclear Legacy

P. Strand

Ladies and gentlemen,

It is my pleasure, on behalf of the Norwegian Radiation Protection Authority, to join representatives from the Russian Federal Medical Biological Agency and North Atlantic Treaty Organisation Science for Peace and Security Programme in wishing you all a warm welcome to this workshop – **Challenges in Radiation Protection and Nuclear Safety Regulation of the Nuclear Legacy** – held in these very fine surroundings here at Ershovo.

As you all are aware, we will be discussing a variety of topics centred on the decommissioning and regulatory control of legacy nuclear sites, seen from an international perspective, during these next 3 days. Here and now, I would like to offer a few words concerning the importance that Norway has placed and continues to place on these topics.

It took the accident at the **Chernobyl** nuclear power plant in 1986 to bring home the possible seriousness of an environmental threat from nuclear accidents close to our borders. Looking back in a positive light, the accident did form the basis for an increased and wide-ranging collaboration between Norway and Russia – our nuclear safety cooperation with Russia derives from the bilateral environmental cooperation agreement established in 1988. Norway and the then Soviet Union signed an agreement on early warning of nuclear accidents and exchange of information on nuclear installations. Norway also collaborated with Russia, Belarus and Ukraine on measures to reduce the impact of the Chernobyl accident on adjacent areas.

From the early 1990s onwards, nuclear safety has been a priority area for Norwegian and Russian authorities. An **expert group** under the Joint Norwegian-Russian Commission on Environment Protection was established in 1992 to investigate whether radioactive waste had been dumped in the Barents and Kara Sea. The expert group has since played a central role in investigations and studies of radioactive pollution, for example at Mayak Production Association, and in the development of cooperation between authorities in the northern areas. The expert group is an excellent example of how Norway and Russia have come together as neighbours to investigate environmental issues that concern us both and work side by side in an effort to find the best practical solutions.

Another example from the early 1990s was from an initiative of the Norwegian government at that time: Norway established a wide-ranging programme to improve protection against **accidents at the Kola nuclear power plant**. In the spring of

Director, Norwegian Radiation Protection Authority, Norway

M.K. Sneve, M.F. Kiselev (eds.) *Challenges in Radiation Protection and Nuclear Safety Regulation of the Nuclear Legacy*,

1994 the Norwegian Government presented a report entitled "Nuclear Activities and Chemical Weapons in Areas Adjacent to Our Northern Borders". This report set the stage for the Government's Action Plan for Nuclear Safety which was then developed in close cooperation between the Norwegian Ministry of Foreign Affairs, Ministry of the Environment, Ministry of Defence, Ministry of Health and Social Affairs, Ministry of Fisheries and the Norwegian Radiation Protection Authority (NRPA). The action plan was initiated in 1995, revised in 1997 and again in 2005, and forms the basis for Norway's nuclear safety collaboration with Russia today. The main aims of the action plan are to protect health, the environment and business activity against radioactive contamination and contamination from chemical weapons in Northwest Russia.

It is important for Norway to ensure that nuclear plants are operated, and radioactive substances handled, in keeping with the highest international standards, and that nuclear materials are properly protected and do not go astray. Norway's support to Russia is helping to reinforce control and supervision with a view to improving the safety of waste storage facilities and nuclear installations and to reducing the risk of future accidents, emissions and radioactive contamination. Norway also wants to contribute to competence transfers that will put Russia in a position to deal with these challenges herself using the best available knowledge and technology, and to contribute to a society that will address these types of problem safely and properly in the future. This requires **risk assessment** studies to be made and the establishment of holistic, cost-effective solutions capable of attracting international support where Russian authorities take on the responsibility of ensuring that this is done in the best possible manner. Norway is seeking to achieve the broadest possible international engagement in this effort, and to spread knowledge of the problems and of what is being done to resolve them. Indeed, awareness of nuclear problems and interest in resolving them has risen appreciably. This has resulted in an extensive international body of rules and guidelines for nuclear activities.

Another important collaboration between Norway and Russia has been the support given for **decommissioning nuclear powered submarines**. Between 2003–2005, Norway has financed the decommissioning of four Victor class nuclear subs at Nerpa and Zvjozdotsjka in Severodvinsk (two type II, one type III & one type I). The last decommissioning project involved close collaboration with Great Britain, reflecting the international efforts from the G-8 nations to help reduce the risk of nuclear incidents at legacy sites in NW Russia.

In our ongoing **cooperation with Russian governmental and regulatory authorities**, Norway is helping Russia to clarify responsibilities and to further develop legislation in the fields of radiation protection, nuclear safety and environmental protection. Wide-ranging cooperation has been initiated with various control and supervisory authorities in Russia. These include Federal Medical Biological Agency (**FMBA**), **GosAtomNadzor** (formerly GAN, now the Federal Technical, Atomic and Environmental Inspectorate), the Health Ministry (**Medbioekstrem**), the Ministry of Atomic Energy (formerly Minatom, now the **Federal Directorate for Atomic Energy**), the **Natural Resources Ministry** and the **Russian Defence Ministry** (radiation protection department).

But we must not forget that the nuclear safety issue is an extensive area in which strong international involvement is needed in order to achieve concrete results. That is why Norway attaches importance to close, frequent contact with Russia, the USA, Canada and the EU in matters related to the northern areas. We hope that these contacts will be strengthened and intensified during this workshop. International cooperation can increase the opportunities for financing projects that are otherwise too large for individual countries to contemplate. Sound coordination means that projects are implemented with the same high requirements in terms of environment, health and safety regardless of which country or institution is responsible. However, to pre-empt poor coordination, responsibility for assigning priorities and for coordinating the international effort must rest with the Russian authorities. At the same time, the various actors need to harmonise their own efforts with the activities of other countries in order to assure synergies and effective resource use.

Our nuclear safety cooperation with Russia brings **regulatory authorities** and specialists from Russia and western countries together. This is important in the effort to assist Russia in its further development of independent and highly qualified supervisory and administrative authorities. At the same time it is important for Russian regulatory authorities to be well prepared to oversee and supervise the clean-up process, supported by international efforts. Right now, our focus has been on helping to **decommission RTGs** collected from NW Russia and working together with the G-8 nations to remediate **the nuclear legacy site at Andreeva Bay**. The NRPA's collaboration with Russian regulatory authorities has shown that rules can be and often need to be improved. Strengthening Russian regulatory authorities will be of major significance. Experience gained by western countries shows that a strong and independent regulatory authority is important to ensure that concrete projects are implemented in a sound manner in environmental and safety terms. The regulatory authorities' position is also crucial to assure sustainable administration of nuclear activities, and ensuring that western assistance will no longer be needed in the longer term.

I hope that this workshop can bring us closer together as partners working to protect the health of our citizens and the environment in which we all live from the threat of radioactive contamination. I can see in the programme there are many interesting presentation, so I'll not hold up the meeting any longer – welcome again and I hope we enjoy a fruitful and invigorating workshop together.

Session I
Nuclear Legacy Challenges

Chairmen: O. Kochetkov, IBPh, Moscow, Russia
and M. Sneve, NRPA, Norway

Issues in Decommissioning and Remediation of Nuclear Legacy Sites

C. Deregel, J.M. Peres, B. Cessac, and P. Francois

Abstract This document is organised in five chapters. The first chapter presents general statements on the different situations to be considered when dealing with nuclear legacy; it can be former nuclear, industrial, research, educational or medical facilities having used nuclear or radioactive material or sites where such facilities were installed.

The second chapter presents the specificities of the dismantling of a nuclear installation and the French regulatory framework to be applied (radiological protection, waste management).

The third chapter deals with the management of radioactive waste and radioactive releases during dismantling of a nuclear installation including conditioning and disposal.

The fourth chapter presents the methodology used in France for the remediation of sites polluted by radioactive substances (guidelines, computer codes).

The fifth chapter presents briefly the organisation of the dismantling site of a French submarine for waste management purposes.

An appendix gives some information on radioactive waste repositories in operation in France.

Keywords Nuclear legacy, dismantling, classes of radio-toxicity, D&D activities, CERISE code, ASTRAL code

1 Nuclear Legacy: Potential Situations

According to the knowledge of already exiting situations all-around the world, one can consider that different situations related to actually out of operation nuclear facilities[1] inherited as "nuclear legacy" may be encountered

Institut de Radioprotection et de Sûreté Nucléaire (IRSN), France

[1] In this document, the term «nuclear facility» refers to nuclear installations as nuclear power plant or research reactors. Other installations with nuclear and/or radiological hazard to be considered are industrial, medical, educational or scientific institutions using or producing nuclear and/or radioactive material.

for their dismantling, including the management of radioactive waste to be produced.

1.1 Former Nuclear Installations Decommissioned Under Regulatory Control

Such situation is the standard situation for former nuclear civil installations in western countries.

In states of the former USSR, a lot of former nuclear installations have been put de facto out of operation without a well organised process due to difficulties for the creation in newly independent states of the necessary regulatory framework after the collapse of the former soviet system.

In such a case, positive elements exist:

- Confident information on the situation of the decommissioned installation.
- Confident information on past activities including incidents or accidents and on existing irradiating or contaminating items is available.
- Negative elements may also exist for installations which have been decommissioned many years ago.
- Disappearance of work craft aware of the operation of the facility (retirement of operators).
- Potentially badly documented situation of radioactive waste, not in line with actual regulations (sorting, classification, characterisation, packaging).

1.2 Former Nuclear Installations Put Out of Operation Without Regulatory Control

It may occur that former nuclear installations have been put out of operation in the past without real regulatory control, even if they are still "under institutional control" or are today without real institutional control ("orphan installations"), as for example former soviet military basis or nuclear ships.[2]
For such installations, negative elements have to be considered:

- Disappearance of work craft aware of the operation of the facility
- Lack of confident information on the actual situation of the installation
- Lack of confident information on radioactive waste and sources left in the installation's territory premises

[2] In this document the term «nuclear ship» refers as well to nuclear powered ships as to floating technical ships supporting nuclear ships.

1.3 Former Industrial, Scientific, Health or Educational Facilities Using Nuclear or Radioactive Material

In such facilities, the presence of radioactive waste and/or orphan radioactive sources must be considered. The radiological remediation of the site will be necessary depending on its future use in order to prevent radiological risks for the public and for the future users of the site (see fourth chapter of the present document).

Several situations may occur.

1.3.1 Industrial Wasteland (Fallow Land) Without Identified Owner

The remediation of such site and/or implementation of protective measures is possible only if the radiological pollution of the site is assessed. In France, historical researches are actually underway for establishing the inventory of such sites.

This inventory allowed to identify, for example, a former industrial site where radium had been used (clock industry); the remediation of the site used for building a school and protective measures (ventilation of the basement of the building for prevention of accumulation of radon) have been performed after assessment of the situation by IRSN experts.

1.3.2 Industrial Wasteland (Fallow Land) Reused for New Activities, the New Owner Is not Aware of the Radiological Contamination of the Site

The situation is similar to the one described above. Remediation measures are necessary as soon as the radiological past of the site has been assessed (national inventory, radiological incident or accident).

The inventory actually underway in France allows to identify former industrial sites where radioactive material had been used (for example radium in the clock industry). On request, IRSN assesses the radiological situation of the site and, if necessary, remediation measures and preventive measures are implemented (removal of "hot spots", of contaminated soil, if buildings already exist, ventilation of their basement for preventing the accumulation of radon).

1.3.3 Industrial Wasteland (Fallow Land) Reused for New Activities, the New Owner Is Aware of the Radiological Contamination of the Site

A good example:

In the years sixties, the former site where laboratories for extraction of radium were installed and used by Marie Curie for her researches up to 1929, has been reused for building a school. The public authorities (National service for radiological protection) recommended protective measures for prevention of risks of irradiation due to the presence of radium remains in the soil (protective sheet of concrete on the soil in the basement of the school, ventilation of the basement).

The better knowledge of the risks induced by radon conducted in the nineties to accurate measurement of radon concentration in the building and in the class-rooms and additional protective measures were performed (ventilation of the class-rooms).

Finally, it was decided to close the school in 1998 in spite of radiological and medical expertises which allowed assessing the very low risks for the actual and past occupants of the school.

Actually the building is closed and its access restricted; the decontamination of the site must be performed before a new use the site.

On the basis of the methodology presented in fourth chapter of the present document and depending on the planned use of the site, the situation will be re-assessed; remediation activities and/or protective measures will be implemented according to the future activities on the site.

2 Decommissioning of a Nuclear Facility

2.1 Specificities

D&D operations are characterised by intensive work inside a nuclear installation with many workers involved, some of them not accustomed to the work in nuclear installations and activities not covered by the standard operation procedures of the nuclear installation.

It means nuclear risks (irradiation by alpha, beta, gamma and neutron emitters, contamination, criticality) and conventional risks (fire, accidents, toxicity, exposure to chemical substances, and so on).

For these reasons, the safety of D&D operations has to be presented in the D&D nuclear safety reports established by the operators in support to the licensing request sent to the nuclear safety authority (NSA) and must be assessed on following aspects:

- Radiological protection (mainly risks of contamination induced by cutting operations)
- Containment of radioactive substances (static or dynamic containment by existing and specific devices (if necessary temporary on place ventilated tenting))
- Environmental protection (radioactive and non-radioactive waste processing, conditioning and temporary storage or disposal, limitation and control of gaseous and liquid effluents releases according to release licenses issued by the nuclear safety authority)
- Fire, chemical and toxicity hazards prevention

When dealing with nuclear legacy installations, the requested studies may be very difficult due to potential uncertainties (see first chapter).

For example, it may be necessary to re-evaluate the studies related to radiological protection, waste, effluents releases if the recovery work of historical radioactive waste reveal unexpected types and/or quantities of waste as alpha emitting waste or fissile substances.

2.2 Regulatory Framework

2.2.1 D&D Activities: French Regulatory Framework

According to the IAEA and ICRP recommendations, the EURATOM and European Community guidelines and regulations, French laws have been passed concerning radiological protection of public and workers, air and water protection, waste management and rules to be followed for pollution-prevention.

Ministerial orders have been written by ministers in charge of health care, environmental protection, industry and defence for the implementation of these laws.

On the basis of these ministerial orders, the operators write technical instructions and specifications to be implemented by the workers involved in the operation of nuclear installations and other facilities dealing with nuclear or radioactive material; these documents are controlled by the NSA.

Concerning the operation and dismantling of nuclear installations, one of the most important documents is the decree of December 31st 1999 issued by the ministry of Industry and Environmental Protection which gives the technical prescriptions to be followed "in order to prevent and limit the harmful effects and external risks induced by the operation of "basic" nuclear installations (INB).[3]

For example, in case of D&D operations, these prescriptions concern following items:

- The prevention of nuclear risks (irradiation, contamination, criticality) in order to protect the environment and the workers.
- The prevention of risks induced by waste; for this purpose the "operating entity" has to identify inside the installation two kinds of zones, namely zones producing conventional (non-radioactive) waste and **zones producing radioactive waste**.
- The establishment and the assessment of a "waste management report" which presents measures taken for waste collection, waste sorting, techniques selected for waste conditioning, interim and final storage of waste and which predicts the yearly trough-out.
- Measures taken for conventional risks prevention (fire, chemical and toxic substances), for noise and vibration prevention, for air and water pollution prevention.

[3] In France, installations containing or using radioactive material are classified as "ICPE". If the amount of radioactive material inside the installation is significant, the ICPE is classified as an "INB". Specific rules in the field of nuclear security and nuclear safety are to be followed for the operation of INBs. An "INB" working for the ministry of Defence is called a "secret INB" (INBS).

The corresponding reports must be assessed and approved by the Nuclear Safety Authority (NSA) prior to D&D operation start-up.

Radiological Protection

The French legislation implements the ICRP 60 recommendations translated in the European directive 96/29.

Based on this legislation, the operators establish technical instructions with "operational" limits lower than the "legal" limits.

For example the operational limits established by major French nuclear operators for their workers (CEA, DGA, DCN)[4] are as follows:

	Regulatory provisions	Technical provisions
Limits	<100 mSv in 5 years And <50 mSv in 1 year	<0.4 mSv/day <1 mSv/week <4 mSv/month
Means of control	Legal passive dosimeters	Active dosimeters (operational dosimetry with direct reading dosimeters by a computerised system (dosiview) and integrated alarm)

When dealing with nuclear legacy installations, the prognoses related to effective doses and to contamination risks may be lower than the reality. It may be necessary to increase the protective measures and to modify working methods and/or to increase the number of workers involved.

Waste Management

The French policy related to waste management for the operation of nuclear facilities, including D&D activities, is based on the principle of "waste zoning" (cf. Section 2.2.1).

The implementation of this principle for DF&D activities aims to facilitate the management of the waste produced and to prevent the risk of dissemination of radioactive waste inside conventional waste.

Two kinds of zones are delimited inside the installation:

- Zones with nuclear waste
- Zones with conventional waste

According to the "principle of precaution", the exemption limits of EURATOM directive 96/29 are not applied in France.

It means that waste produced in a zone producing nuclear waste cannot be considered as "conventional" even if the activity after measuring is under the detection limit and under the exemption limit of the EURATOM directive (IAEA exemption levels).

[4] CEA: Atomic Energy Commissariat, DGA : National Armament Agency, DCN: Military Ships Building Agency.

It is a French specificity and it limits the possibility of recycling very low activity radioactive waste, namely scrap metal and concrete rubble (obligation to send such waste to a disposal dedicated to very low level waste or re-use in the nuclear industry for example scrap metal melted and used for manufacturing shielding integrated in waste packages).

The practical implementation of waste zoning by French nuclear operators is made on the basis of the knowledge of the installation and of its actual and past operation (consideration of past incidents and accidents); on this basis a *theoretical delimitation* of the waste zones is made as follows:

1. Zones with nuclear waste: such zone is called "*contaminating zone*" and can produce radioactive waste (HLW, LLW, VLLW) due to the actual presence of radioactive material and/or contamination.
2. Zones with conventional waste: two kinds of zones are delimitated:
   (a) *Non-contaminating zone*: in such a zone, radioactive material was present in the past but today there are no risks of contamination (cleaning and decontaminating activities performed).
   (b) *Zone without added radioactivity*: in such a zone, radioactive material was never present.

The theoretical delimitation of waste zones is confirmed by accurate measuring of dose-rate, activity and surface contamination with very low detection limits much lower than IAEA exemption levels.

The limits of each zone are clearly marked and access doors are installed for the passage from one zone to the other.

Conventional and radioactive waste are manages by two completely separated channels.

The waste zoning may be modified during the D&D activities, for example cutting of an internally contaminated tank located inside a non-contaminating zone.

When dealing with a nuclear legacy installation, the definition of the theoretical zoning may be very difficult, notably for the definition of zones without added radioactivity. Tremendous measurement operations may be necessary for confirmation of the theoretical zoning (delimitation of the non-contaminating zones) and probably big changes will occur during the D&D activities (uncertainties on the real situation). High quantities of very low level waste may be produced if it is not possible to identify non-contaminating zones and zones without added radioactivity.

3 Waste and Effluents Management During French D&D Activities

3.1 Management of Liquid and Gaseous Effluents

There are two options for the management of gaseous and liquid radioactive effluents:

1. Immediate or delayed (use of the radioactive decay properties) release in the environment
2. Temporary storage before treatment and final conditioning for long term storage or disposal depending on the existing disposal facilities

Ministerial orders lay down the conditions for release of effluents in the environment:

1. License issued by the nuclear safety authority on the basis of a nuclear safety report containing an environmental impact study of the releases and the technical prescriptions proposed for the releases of radioactive and toxic effluents (concentration limits, maximal yearly release, control of the activity released, environmental monitoring)
2. Yearly report about the activity released and results of environmental monitoring
3. Immediate declaration in case of non-controlled releases or releases exceeding the limits fixed by the license

3.2 *Management of Solid Waste*

Non-radioactive Waste

The non-radioactive waste is managed according to the regulatory framework in force (toxic waste, asbestos). Licensed dumping facilities may be used for disposal (Installations classified for environmental protection ICPE).

3.2.2 Radioactive Waste

Present Situation in France

The French nuclear industry in France has gathered a lot of experience in operation and dismantling of nuclear installations (NPPs, laboratories, nuclear submarines) and managing of radioactive waste (treatment by incineration and melting and conditioning (SOCODEI activities), disposal (ANDRA activities)).

According to international standards, France has established a comprehensive set of rules concerning the operation and the dismantling of nuclear installations.

These rules contain a lot of provisions concerning radioactive and non radioactive waste management (waste zoning inside nuclear installations, acceptance criteria for radioactive waste packages in the French disposal centres for low-level and very-low-level waste, obligation for waste producers to reduce the volume of waste produced and to recycle with the aim to minimize the quantity of ultimate waste to be disposed).

As of the early 1990s, the national French electricity producer EDF and the French Company COGEMA (reprocessing of spent nuclear fuel) clearly demonstrated their determination to acquire the industrial means necessary to face the challenges raised by the management of waste produced by operation and dismantling of nuclear installations with the creation of SOCODEI (French Society for industrial Waste treatment).

SOCODEI was assigned the missions of designing, building, financing and operating the installations for the treatment of low-level and intermediate-level radioactive waste.

Accordingly, since 1999, SOCODEI has operated the basic nuclear installation CENTRACO (Nuclear Waste Treatment and Conditioning Plant) located near the Marcoule site (Rhone valley). CENTRACO contains, on a single site, an incineration unit for burning liquid and solid waste, a melting facility for metallic waste and a recycling unit for scrap metal.

The institute for Radiological Protection and Nuclear Safety (IRSN), in support to French and foreign Nuclear Safety Authorities (NSA), assesses nuclear safety reports established by operators in support to licensing request for operation and dismantling of nuclear installations.

For example, IRSN assessed nuclear safety reports concerning decommissioning operations of following installations:

- French NPP's (graphite-gas reactors …)
- French nuclear submarines
- Foreign NPP's (feasibility studies for the decommissioning of KOZLODUY 1 and 2 in Bulgaria, INGALINA in Lithuania)

as well as the nuclear safety reports for the licensing of the CENTRACO facility.

Management of Radioactive Waste Generated by D&D Activities

According to the regulatory framework and to existing licensed treatment and disposal facilities for radioactive waste in France, following options are available:

- Disposal in ANDRA facilities (see annexe) after sorting, treatment (volume reduction when possible by compacting, incineration or melting) and conditioning accordingly to acceptance criteria related to the repository:
  - Centre de Stockage de l'Aube (CSA) for low and intermediate short lived waste
  - Centre de Stockage TFA de Morvilliers (CSTFA) in operation since 2002 for very low level waste (the first facility of this kind in the world)
- Temporary storage in agreed installations for the other categories of waste (studies and experiments are underway for transmutation or disposal of high activity long lived waste, for disposal of radium bearing waste and for graphite waste)

The annexe presents some features of the two repositories actually in operation in France.

4 Remediation of Sites Polluted by Radioactive Substances

4.1 Introduction

One of the problems raised concerning the impact of sites polluted by radioactive substances is the real or potential consequences on human health of this radioactive pollution.

Depending on the nature of the radioactive pollution (sealed source, dissemination of radioactive substances in the soil, in water, in the air, fixed or loose contamination of surfaces), it is necessary to evaluate the consequences in terms of equivalent biologic dose for human beings due to external and internal irradiation induced by all potential ways of aggression of human beings (direct exposure, ingestion of contaminated food, ...).

In case of doses exceeding the accepted norms, it is necessary to take measures in order to achieve an acceptable detriment level for human beings.

A lot of difficulties exist when somebody tries to assess the environmental impact of a site polluted by radioactive substances (difficulty to characterize the source, to assess the release of the radioactive substances in the geosphere, to predict their diffusion in the different medias and their transfer to the food-chain of human beings and; consequently, to asses the effectiveness of the measures implemented to reduce the nuisance of this use).

These difficulties are summarized on the following scheme (Fig. 1):

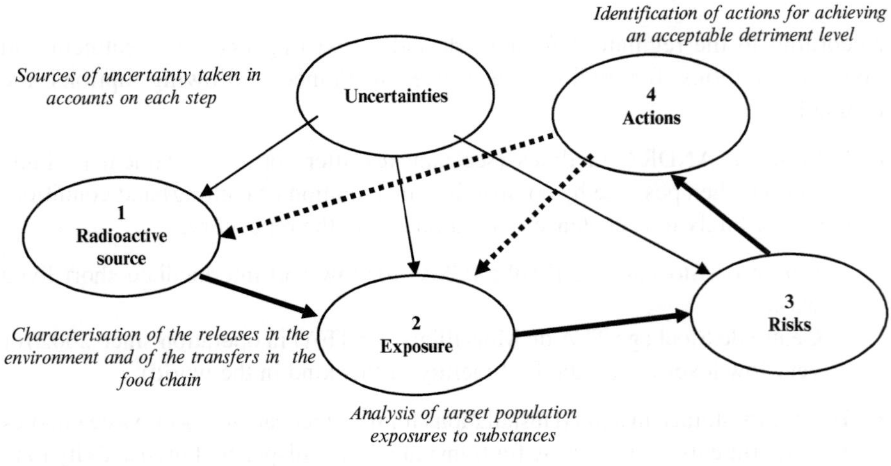

Fig. 1 Relationship between actions and risks

4.2 *Expertise and Research of the IRSN*

In the field of polluted sites and soils, the IRSN carries out the following activities:

- Radiological expertise on the sites contaminated by radioactive substances: characterization of the sites and evaluation of their dosimetric impact on all the populations concerned, including the workers, and on the environment
- Security of the sites, development of the protocols of cleaning and technical aid to the administrative follow-up of the works
- Control of work and work of building site of drainage work. Reception of work and checking of conformity to the remediation objectives
- Research in support with these expertises, for a better comprehension of the phenomena, which lead to the dissemination of the radionuclides in the environment

(Fore more details, see the IRSN web-sites www.irsn.org and www.irsn.org/net-science/)

4.3 *Methodology for the Management of Industrial Sites Contaminated with Radionuclides in France*

4.3.1 A Methodologic Guide

The French authorities assigned the task of preparing a guide for sites potentially contaminated by radioactive substances to the Nuclear Safety and Radiological Protection Institute (IRSN). For the sake of consistency, they requested that the guide be based, as far as possible, on the approach adopted for sites contaminated by chemicals. The study completed by the end of 1999 has been approved by the two ministries (Health and Environment) that commissioned it and made public in 2000.

The aim of the guide is to provide an operational framework for the management of radioactively contaminated sites, which will replace the current case-by-case approach by a set of recognized procedures that will ensure the "trace ability" of the whole process from assessment to decision. It will provide a system of reference for all the stakeholders involved and will permit dialogue on a common basis.

This guide deals with the various situations that may be met in France when rehabilitating industrial (non nuclear) sites (potentially) contaminated with radioactive substances. These sites are defined as the ones where the ground or buildings have been contaminated by activities involving radioactive substances, which have taken place either on the site itself or nearby.

In principle, various forms of "contamination" are to be found at these sites:

- The soil and buildings at the site may have been contaminated by radioactive substances involved directly, or as by-products, in manufacturing processes and in research work. In soil contamination, these substances are present in various

concentrations. Contamination of buildings occurs in two forms: loose contamination and fixed contamination.
- Sealed or unsealed sources may have been handled and left on site, in such a way that the radioactive substances are still contained in their packaging or have migrated following degradation of the packaging.

The guide presents an approach involving several stages:

- Removal of the doubt
- Pre-diagnosis
- Initial diagnosis
- Simplified risk assessment
- Detailed risk assessment
- Assistance in selecting the remediation strategy for a given use

The guide outlines the criteria which enable the assessment sequence to be interrupted and the appropriate decisions to be taken. For example, one can stop at the stage of the simplified risk study when the site is small and if it is relatively easy to remove and store the contaminated soil. The selection of the appropriate strategy presupposes the identification of several alternate options which must be characterized in terms of reduction of dosimetric impact, reduction of contamination, costs and associated nuisances. The choice of a remediation strategy requires the close involvement of the stakeholders.

The radiological aspect is generally only one of the elements of the choice and conditions have to be created to enable the stakeholders to discuss all the aspects relevant to the specific context of the site.

It may not be necessary to implement all stages. In particular, the guide distinguishes between contaminated soils and contaminated buildings, where the approach is applied in different ways. In any event, the assessment effort, which is often very costly and time-consuming, should take into account the characteristics of the situation encountered e.g. level of contamination, future use of the site (sensitive use as for example residential area, non sensitive use as for example car-park), and so on.

Example of Process: The Simplified Risk Assessment

The simplified risk assessment (SRA) involves the calculation of the potential dosimetric impact associated with various scenarios for the use of the site and buildings, based on the results of radioactivity measurements of the soil and buildings at the site.

It takes in account the results of the initial diagnosis (historical analysis, vulnerability of the environment due to the geological and geographical characteristics (consistency of the soil, ground water, surface water, air), radiological characteristics (mapping of the surface, first subsurface studies, measurements of radioactivity in water, crops and animals which may enter the human food chain).

In order to facilitate the calculation of the dosimetric impact, generic scenarios (home, primary school, offices; market garden, car park) have been prepared and then evaluated using a dose calculation model. This model determines the individual effective dose in mSv/year associated with soil contamination that is incurred per unit of specific activity (1 Bq/g of soil) for the radionuclides likely to be found at contaminated sites. The generic scenarios incorporate simple assumptions that prudentially over-estimate the dose impact.

The following tables give some examples of the results obtained:

	Annual effective dose (mSv/year) for 1 Bq/g of each radionuclide or decay chain						
	Residential area	Agricultural land	Primary school	Office	Parking area	Waste land	Building site
^{232}TH+	0,640	1,160	0,230	0,064	0,016	0,220	0,240
^{238}U+	0,490	0,805	0,170	0,045	0,011	0,175	0,170

	Activity (Bq/g) for a dose of 1 mSv/year of each radionuclide or decay chain						
	Residential area	Agricultural land	Primary school	Office	Parking area	Waste land	Building site
^{232}TH+	1,6	0,9	4,4	16	62	4,5	4,2
^{238}U+	2,0	1,2	5,9	22,4	89	5,7	5,9

Public authorities lay down the acceptable limits for each category of use (selection level or SL) in mSv/year (generally a fraction of 1 mSv/year) and this selection level is used to decide if remediation work has to be carried out and/or if limitations in the use of the site or of some parts of it must be established.

The calculations made allow determining the effective dose DJ corresponding to the different "generic uses" of the site considered.

Depending on the comparison between DJ and SL, and on the sensitivity of the considered use (for example the use as primary school is sensitive) different decisions can be taken as illustrated on the diagram on next page.

Summing Up

One of the main features of the assessment approach presented here, which resembles that used in the guide on chemical substances, is its sequential nature.

Bearing in mind the cost and length of the studies involved in implementing the approach, the latter must be adapted to specific circumstances if it is to be effective. It will not always be necessary to carry out every stage of the assessment.

If a detailed risk assessment turns out to be necessary, it may be useful to divide the site into various sectors if it is large and has a highly non-uniform contamination. This division could be made on the basis of the remediation techniques that should be applied to each sector.

Finding the appropriate strategy means identifying several options which then have to be described in terms of various factors such as reduction in radiological impact, cost, reversibility, service life of the remediation measures, need for institutional monitoring and maintenance of the site, and so on.

Regardless of whether this investigation is accompanied by a discussion on selecting a site use, consultations involving all the stakeholders should take place. The scale of these consultations will, of course, depend on the context.

Generally speaking, managing a contaminated site is often a lengthy process, which may extend over several years, from the time contamination is found until remediation has been completed.

The following diagram (Fig. 2) presents the simplified risk assessment for a contaminated soil:

4.3.2 Some Methods and Tools Available

Calculation of Doses

CERISE Calculation Code

The CERISE code (Code for Individual Radiological evaluation for activities in firms and in open air) allows calculating the exposure for preplanned scenarios (generic scenarios as fallow land, building place, residential area, primary school, offices, market gardener, car park) considering:

- The external exposure (β, γ) due to simple geometric forms (point sources, cube, sphere, infinite flat surface, …
- Mixed exposures (inside a house with all ways of aggression and atmospheric emission (external exposure and inhalation)
- Results are given in Sievert/year/man of by activity

Ciblex Programme

The characterization of situations that may lead to an exposure of the persons living on a contaminated site or within its immediate environment needs to have some knowledge of the routes of exposure of the individuals concerned, but also to be able to perceive the parameters that best characterize the behaviour of these individuals in their daily life.

In order to characterize the French population related to the management of contaminated sites, the Environment and Energy Management Agency (ADEME) and the Institute for Radiological protection and Nuclear Safety(IRSN) have established a study agreement named the « Ciblex Study », which aims to establish a data base usable for sanitary risks assessment in case of an exposure to radioactive substances.

ASTRAL Calculation Code

For the assessment of the consequences of the release of radioactive substances outside a nuclear installations in case of an emergency, the ASTRAL code (Technical Assistance in Radioprotection after an accident) has been developed in

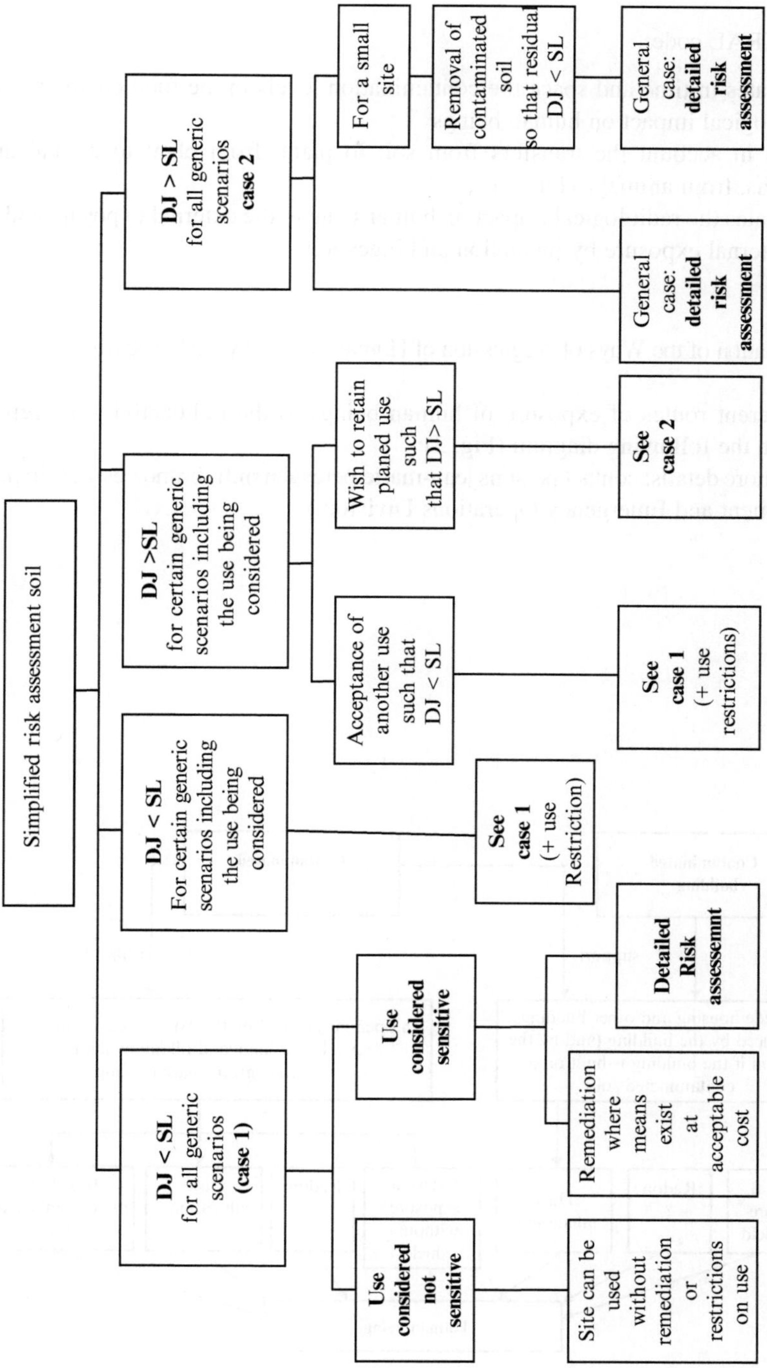

Fig. 2 Simplified risk assessment. (For more details: contact persons jean-marie.peres@irsn.fr, bruno.cessac@irsn.fr Environment and Emergency Operations Division)

order to quantify the transfer of radioactive elements in agricultural areas after an accident.
The ASTRAL code:

- Evaluates in time and space the contamination levels in the food chain and the radiological impact on human beings
- Takes in account the transfers from soil to plant, from plant to animal and humans, from animal to humans
- Evaluates the radiological impact on humans due to the external exposure and to the internal exposure by inhalation and ingestion

Determination of the Ways of Aggression of Human Beings by Radioactivity

The different routes of exposure of human beings to the radioactivity are represented on the following diagram (Fig. 3):

(For more details: contact persons jean-marie.peres@irsn.fr, bruno.cessac@irsn.fr, Environment and Emergency Operations Division.)

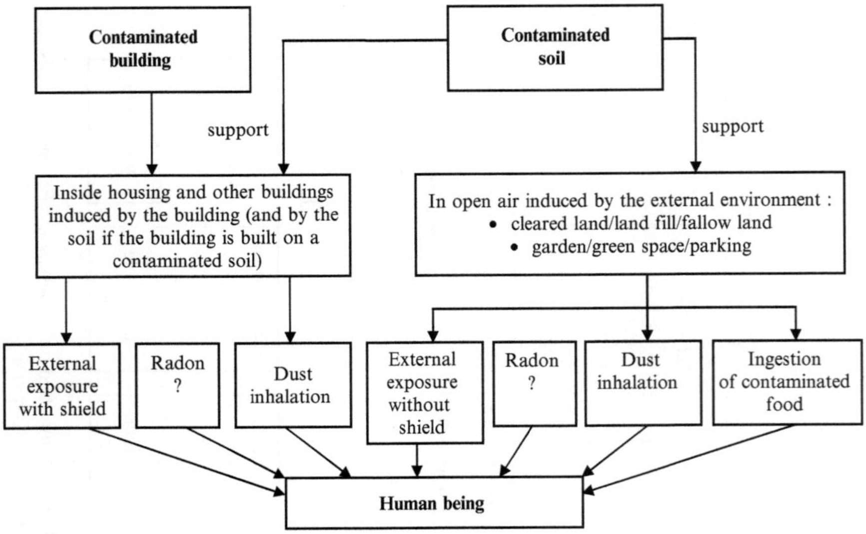

Fig. 3 Summary of exposure modes

5 Example: Decommissioning of French Nuclear Submarines

5.1 Implementation of the French Regulation Concerning Waste Zoning for the Decommissioning of Submarines

The implementation of the waste zoning by the operators on board a submarine under decommissioning is made in following manners, taking in consideration the past activities on board the submarine (including incidents which occurred) and the actual activities:

1. **Zones with conventional waste**: the operator delimits to kinds of zones:
   (a) **Non-contaminating zone**: it means a zone where radioactive material and/or radioactive contamination was present in the past but which has been "cleaned" (removal of radioactive material, decontamination), for example reactor's auxiliary rooms
   (b) **Zone without added contamination**: it means a zone where radioactive material was never present, for example operational centre, living quarters of the crew
2. **Zones with radioactive waste**: the operator delimits on board "**contaminating zones**", it means zones where the production of radioactive waste with high, low or very low activity is possible due to the actual presence of radioactive material (reactor's compartment, temporary zones when cutting for example a pipe which contained primary coolant inside a zone classified as "non-contaminating zone")

The following schemes (Fig. 4) summarize the organisation used for waste management and environmental protection:

This organisation aims to prevent a dissemination of contamination and a mix between conventional and radioactive waste (separate channels for control, conditioning and transport).

Annexe: Radioactive Waste Disposal Facilities in Operation in France

1 Disposal for Short and Intermediate Short Lived Waste

The "Centre de Stockage de l'Aube" (CSA) is based on the concept of surface disposal in vaults (400 bunkers in concrete) installed above the table of ground waters. In entered in operation in 1992 and its capacity (1 million m^3) allows to receive radioactive waste produced in France during 50 years.

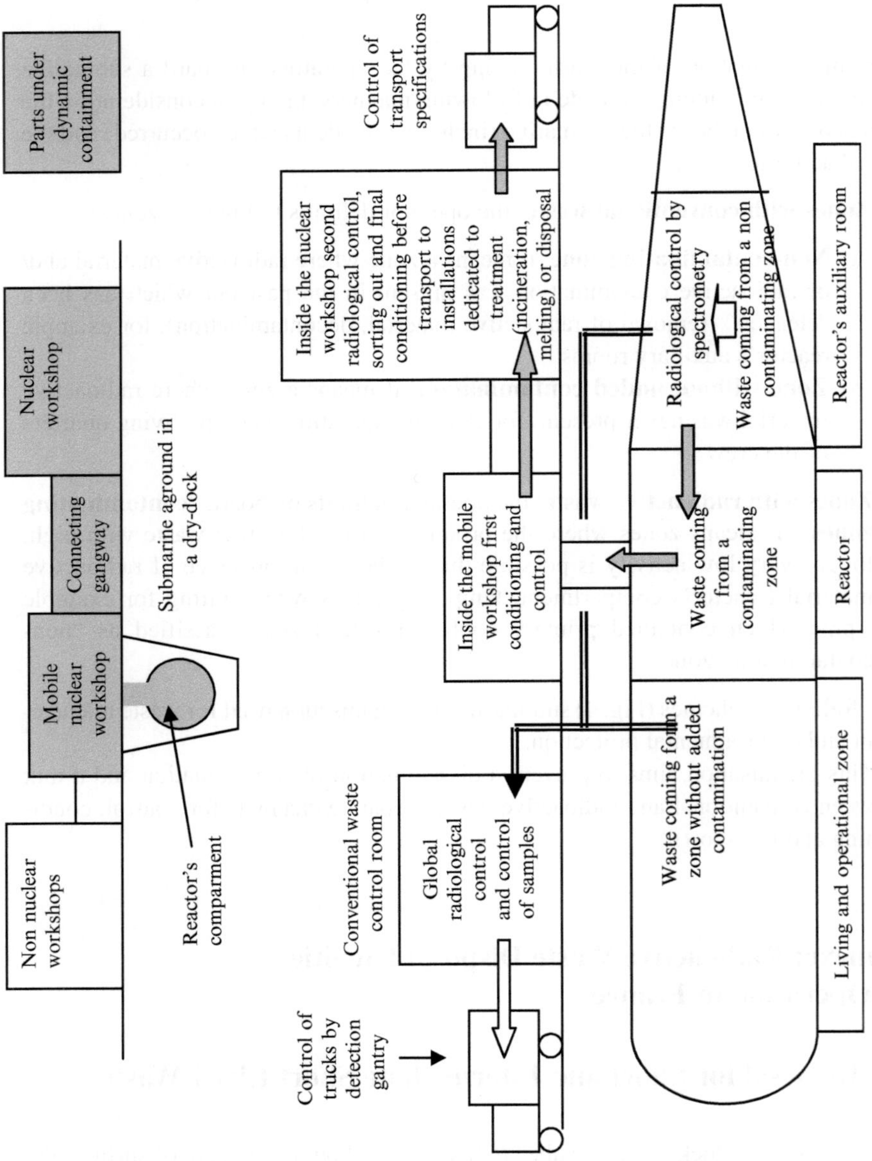

Fig. 4 Diagram of waste management activities during submarine decommissioning

Acceptance criteria related to radionuclide, activity, embedding or not and packages as well as the rules for the operation and the monitoring of the facility and its environment have been established by ANDRA (Figs. 5 and 6), approved by the Nuclear safety Authority and are recalled in the ministerial order issuing the permit for the operation of the facility which is classified as Nuclear Basic Installation.

The main principal is that after 300 years, the radioactivity induced by the disposed waste shall be equivalent to the natural radioactivity.

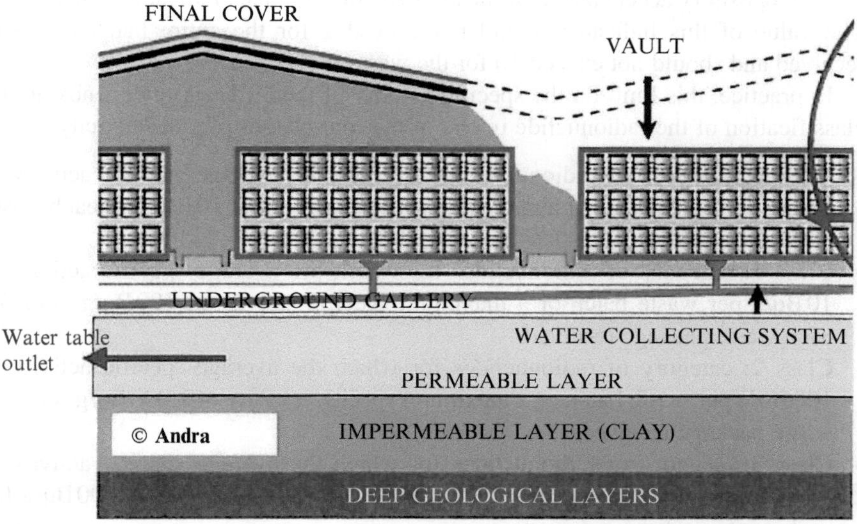

Fig. 5 Scheme of the facility

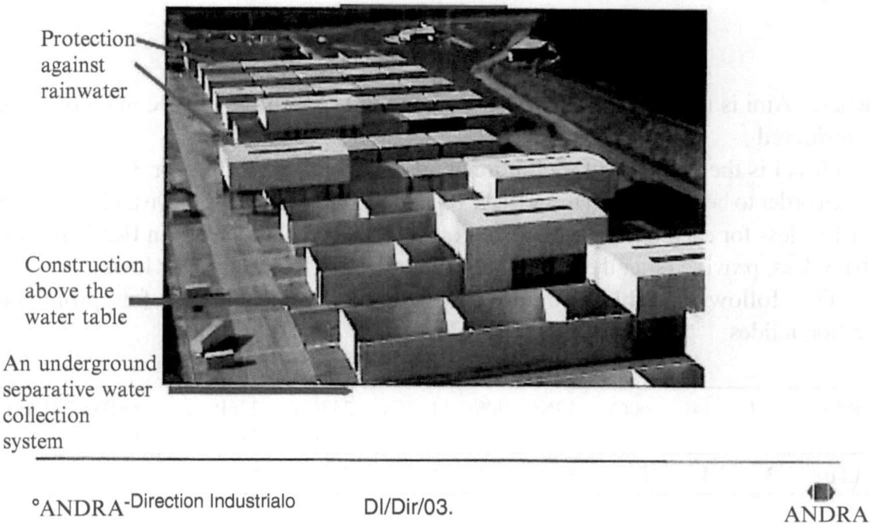

Fig. 6 Illustration of waste compartments (For additional information please consult the ANDRA website www.andra.fr.)

2 Disposal for Very Low Level Waste

The "Centre de Stockage de Morviliers" (CSTFA) is based on the concept of disposal in trenches dug in an adapted soil (Fig. 7). It entered in operation in 2003 and its capacity (650,000m^3) is planned for receiving the French VLLW produced during 30 years.

The principle used for defining the acceptance criteria is that the activity of the disposed waste after 30 years shall be equivalent to the natural radioactivity.

The repository acceptance criteria are based on an impact indicator called IRAS. The value of this indicator should not exceed 1 for the entire batch of waste received and should not exceed 10 for the waste.

In practice, this limit on the specific activity of the VLL waste depends on the classification of the radionuclide in one of the four classes of radio-toxicity:

- Class 0: category of radionuclides for which the average specific activity is 1 Bq/g per waste batch or a maximum specific activity of 10 Bq/g for each waste package received.
- Class 1: category of radionuclides for which the average specific activity is 10 Bq/g per waste batch or a maximum specific activity of 100 Bq/g pour for each waste package received.
- Class 2: category of radionuclides for which the average specific activity is 100 Bq/g per waste batch or a maximum specific activity of 1,000 Bq/g for each waste package received.
- Class 3: category of radionuclides for which the average specific activity is 1,000 Bq/g per waste batch or a maximum specific activity of 10,000 Bq/g for each waste package received.

To determine the acceptability of a batch of waste, the repository radiological acceptance index (IRAS) is defined as follows:

$$IRAS = \Sigma\,(Ami/10\,class\,i)$$

where: Ami is the specific activity of radionuclide i (in Bq/g) in the mass of waste considered

class i is the class N° of the radionuclide i considered (0, 1, 2 or 3).

In order to be accepted in the VLLW repository, the waste must have an IRAS index of 1 or less for each batch. A waste package in this batch may have an IRAS index of 10 or less, provided that the average index for the batch remains 1 or lower.

The following table provides the corresponding classes for prominent radionuclides:

RNi	3H	14C	60Co	63Ni	90Sr	137Cs	232U à 238U	236Pu up to 240Pu, 241Am 242Pu, 244Pu
Class	3	3	1	3	3	1	2	1

The reference activity for cobalt 60 and caesium 137 is 10Bq/g, while it is 100Bq/g for the isotopes of uranium and 1,000Bq/g for tritium.

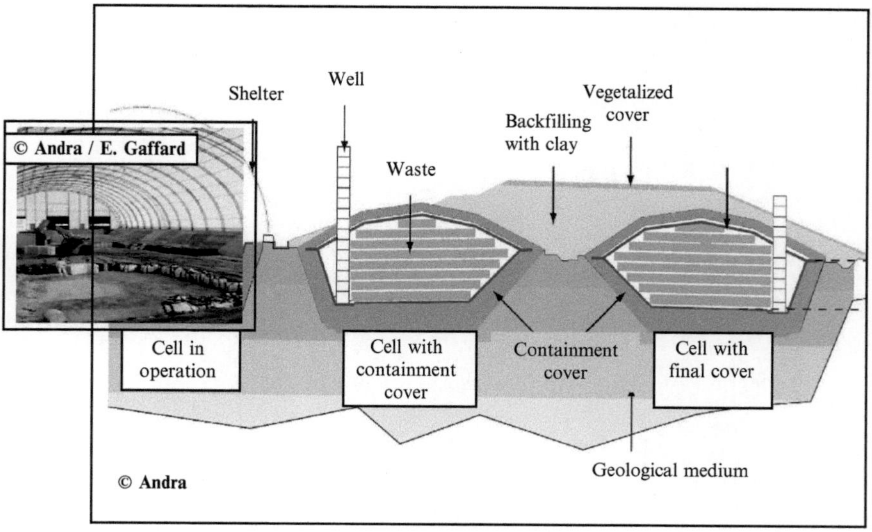

Fig. 7 Scheme of the conception of the CSTFA

Unified State System of Management Spent Nuclear Fuel and Radioactive Waste. Conceptual Approaches and Generation Principles

I.V. Gusakov-Stanyukovich

Abstract Provision of extended reproduction of atomic branch manufactures is the main objective of the RF atomic branch development program, approved by the RF President in June of 2006. One task to be solved for this goal reaching is arrangement of the state safety guarantee system at atomic energy use, in particular, by means of generation of the unified state system of SNF and RW management. This system is summarised in this paper.

Keywords SNF, RW, “Mayak”, EGSO

Now, there are a number of reasons to consider the unified **state system of SNF and RW management** (hereinafter referred to as EGSO) under generation as an infrastructural design with respect to the program of atomic power industrial combine in the Russian Federation. At the same time, EGSO must ensure the “historical legacy” problem solving. We mean here decommissioning of nuclear and radiation hazardous facilities (NRHF) and management of SNF and RW accumulated.

The following factors were taken into account as prerequisites for EGSO generation in a form of the unified system:

- Atomic energy use relates to SNF and RW generation, cost for management of which make significant contribution into expenses for production (operations, services etc).
- It is impossible (not cost effective) to arrange a full cycle of SNF and RW management within a single operator because of high cost of reprocessing the necessity to provide long-term RW isolation.
- The necessity of the state EGSO status at the present stage is caused by:
- Obligations taken by the Russian Federation within «Joint convention on safe SNF and RW management» (Vienna, 5 September, 1997), according to which the State is the final responsible for safe SNF and RW management.
- The state status of the program of atomic power industrial combine development.

Federal Atomic Energy Agency (Rosatom), Russia

- Recognized State responsibility for solving problems connected with "historical" SNF and RW, decommissed nuclear and radiation hazardous facilities, which are the legacy of military programs and planned economic activity (Federal target program «Nuclear and radiation safety assurance for 2008 and for the period up to 2015»).

The main infrastructure elements of EGSO for the long-term perspective are:

- Federal facility (facilities) for long-lived RW disposal into deep geological formations
- Regional (local) disposal facilities for short-lived RW and/or very low level RW
- Federal storage facility (facilities) for SNF of WWER-1000 type reactors, RBMK and non-standard SNF

Flowchart 1 (Fig. 1) Shows The Strategy Of RW Management In Russian Federation.

Ability is under consideration of RW shallow disposal facilities arrangement to serve as regional units in the European region of Russian Federation, Southern-Ural region, East Siberia. Local units for RW disposal can be arranged on 35 Rosatom's and Rosstroy's sites.

Flowchart 2 (Fig. 2) shows the strategy of SNF management infrastructure arrangement in Russian Federation.

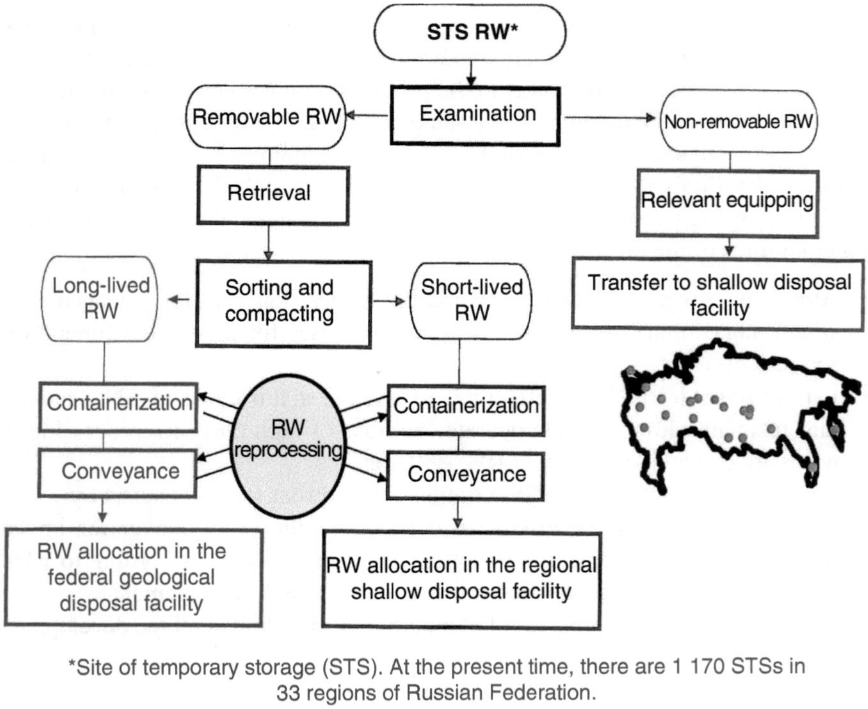

Fig. 1 The Strategy of RW Management in Russian Federation

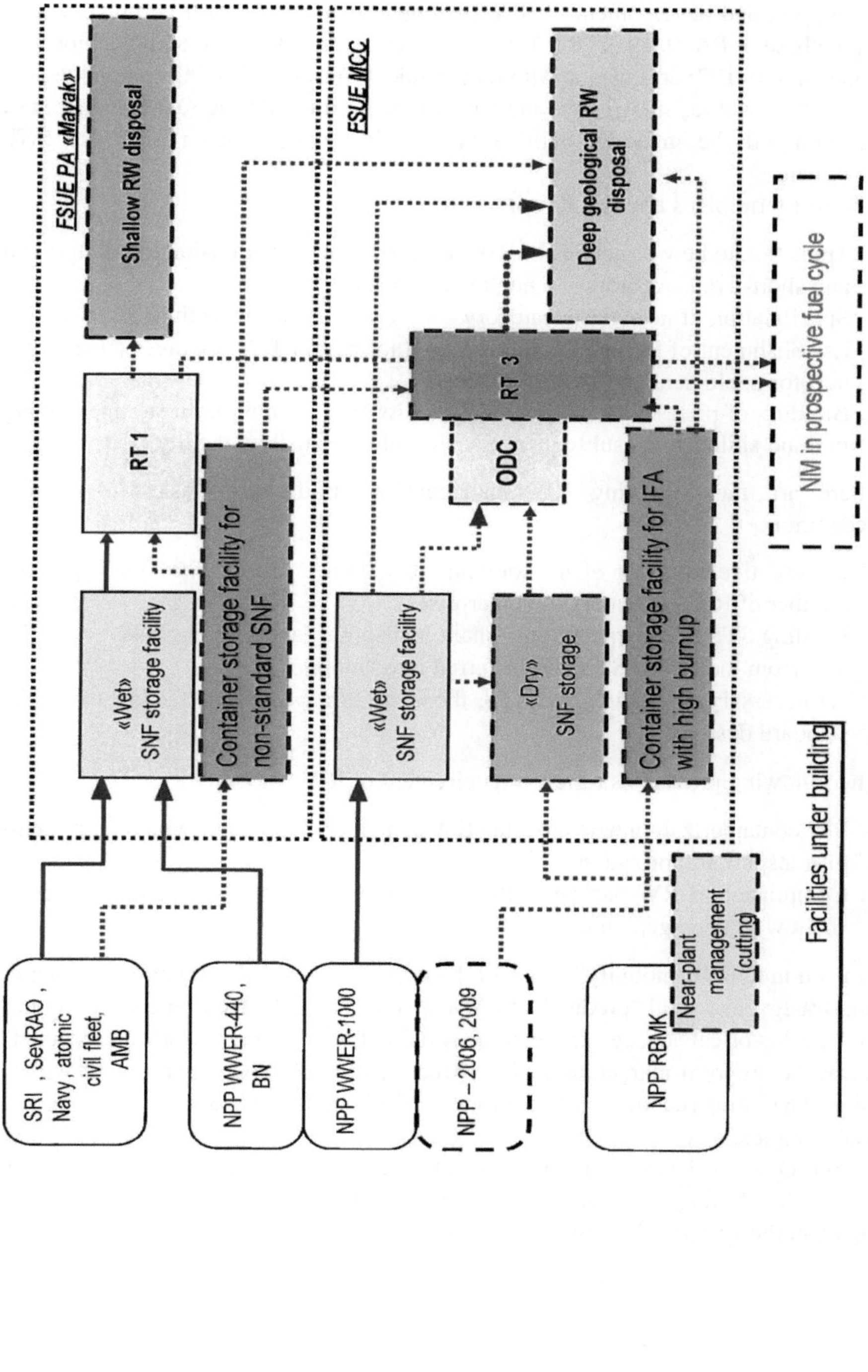

Fig. 2 Strategy of SNF Management Infrastructure Arrangement in Russian Federation

In the nearest future, "dry" storage facility, which is under construction currently in Krasnoyarsk area, will be the main infrastructural element, providing NPP SNF management. Arrangement of SNF storage facilities using container technologies is possible.

Support and development of SNF treatment technologies will be implemented on the base of PA "Mayak" RT-1 plant, as well as that of experimental demonstration centre (ODC) arranged at Mining Chemical Combine. Having in mind results of work as well as knowledge and technologies gained during ODC operation, a decision will be made to build a large-scale next-generation plant for SNF treatment.

Short-term plans are as follows:

- Transition to new principles of RW categorization – subdivision into long-lived and short-lived, according to nuclide composition
- Specification of new waste category – very low level (low active) RW
- Establishment of technical requirements (norms) for RW packages (final product) for geological and shallow disposal
- Building of pilot (standard) regional facility for RW long-term storage, having in mind ability of its subsequent transfer into the shallow facility

There are the following RW management middle-term tasks in Russian Federation:

- Quantitative reduction of sites of temporary storage (now 1 170 STSs) up to the number of RW generators (69 enterprises)
- Existing STS either transfer into shallow disposal facilities, or close
- RW from the closed STS is transferred into "mobile" state
- As necessary, instead closed STS, those are arranged at the facility under the standard design

The following provisions reflect a requirement of RW "mobility":

- RW containerization with ensuring RW container compliance with requirements of transport authorization
- Compliance of RW package with acceptance criteria with intension of further shallow or deep geological disposal

Transition to RW "mobility" is one of the important elements of both EGSO building strategy, as a final "product" of RW management at the facilities, and programs of "the historical legacy" mitigation. Availability of demand for "the product" forms the proposal-market, in the infrastructure of which some investments will be done. The infrastructure will begin to work for benefit of the atomic power industrial complex.

EGSO must become an element of the State atomic energy corporation "Rosatom" developed in Russian Federation. Figure 3 shows EGSO building principles in the system of the State atomic energy corporation.

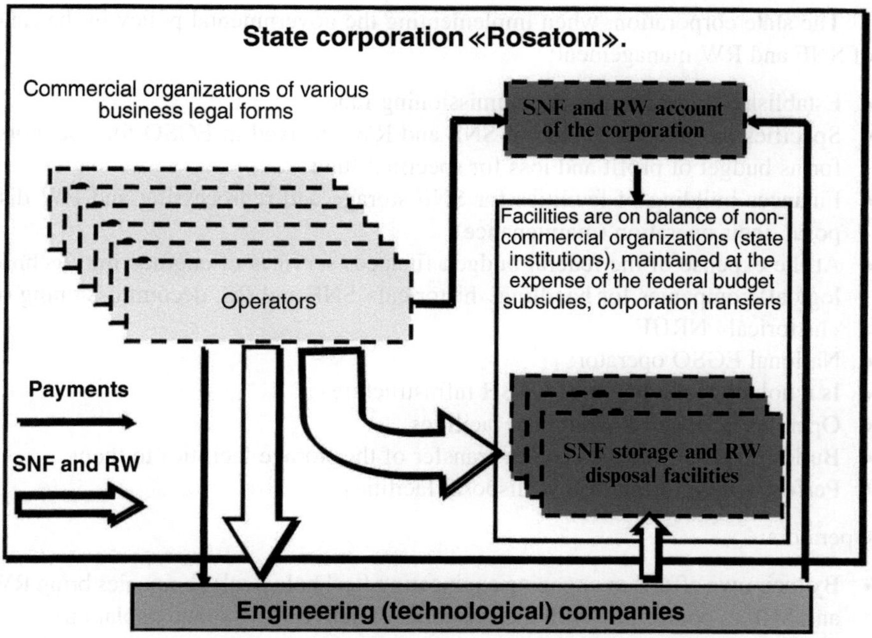

Fig. 3 ROSATOM and EGSO interactions

EGSO will be based of the following main principles:

«Contaminator pays». Operator, when generating waste, carries a burden of the financial responsibility for their management.

«Pay and forget». Operator's responsibility for SNF and RW management terminates after:

- SNF and RW putting into compliance with EGSO standards
- Appropriate payment as well as SNF and RW transfer to the special EGSO organization (organizations)
- Rights of SNF and RW ownership in EGSO system are considered in the following aspect:
- Since a time of generation, SNF and RW are the operator's property.
- Appropriate payment as well as SNF and RW transfer to EGSO national operator means also transfer of their property rights.
- Rights of SNF and RW ownership, being transferred to engineering (technological) companies, are specified in the contract (treaty) and established for the party, which is responsible for transfer of treatment products to EGSO national operator.

The main principles of the responsibilities division at SNF and RW management, accounted in EGSO model under consideration, will be as follows:

The state corporation, when implementing the governmental policy in the field of SNF and RW management:

- Establishes and manages decommissioning fund
- Specifies the payment value of SNF and RW received in EGSO for operators, forms budget of profit and loss for specified funds
- Finances building of facilities for SNF storage and reprocessing and RW disposal, their operation (maintenance)
- At the expense of the federal budget, finances services of engineering (technological) companies for handling «historical» SNF and RS, decommissioning of «historical» NRHF
- National EGSO operator:
- Is a holder of the books of USSH infrastructure
- Operates SNF and RW storage facilities
- Builds disposal facilities and/or transfer of the storage facilities to them
- Performs monitoring of RW disposal facilities

Operators:

- By their own efforts or engaging engineering (technological) companies bring RW and SNF to conformity with the standards of delivery to storage (isolation)
- Based on prices specified by the State Corporation deliver SNF and RW to EGSO national operator
- Assign to the fund means for decommissioning of NRHF
- Pay services of engineering (technological) companies for reprocessing SNF and RW if it's cost-effective

Engineering (technological) companies performs:

- Development of RW management technologies (performance of research and development)
- Works (services) for the stages of handling RW and SNF
- Transportation of SNF and RW from the operator areas to the SNF storage and RW disposal facilities

In the nearest perspective, operation of storage facilities on the base of current expenses financing is transferred (can be transferred) to operators.

Engineering (technological) companies of all ownership types are admitted to works if they have activity license.

The following stages of EGSO of SNF and RW in the State corporation system of atomic branch are:

Scientific methodic stage including:

- Formalization (standardization) of requirements to RW for receipt to final isolation
- Development of standard programs for management of «historical» SNF and RW
- Justification of the locations and making decision for SNF, RW long-term storage facilities, RW disposal facilities

Organizational-preparatory stage, including:

- Centralization of calculation and justification of payments into SNF and RW account of the State Corporation
- Development of regulations for funding maintenance of «historical» SNF and RW, payment of the works on their conditioning and delivery to EGSO
- Start of National EGSO national operator work
- Start of practical works, including:
- Implementation of the pilot design of the regional RW disposal facility
- Making demand for works (services) for RW management and decommissioning of NRHF based on approved standards and government work as related to «historical» RW
- Consolidation on the balance of the national operator of SNF and RW storage infrastructure

EGSO stable operation stage:

- Receipt of SNF and RW from operators for a fee
- Investment of spare cash into EGSO infrastructure arrangement

Nuclear Legacy Problems and Their Solutions Within the Federal Target Program «Nuclear and Radiation Safety Assurance for 2008 and for the Period Till 2015»

I.I. Linge

Abstract This paper analyses problems of nuclear legacy caused by different reasons: technological, economic, etc. These challenges relate to SNF and RW management, nuclear submarines dismantlement, remediation of shore territories, and decommissioning of nuclear and radiation hazardous facilities. The advantages are discussed of the programs and target method applied for the mentioned problem solving.

Keywords SNF, RW, decommissioning, Novovoronezh NPP, radiation hazardous facilities, federal target program (FTP)

The atomic industrial branch originated in 1940s. The nuclear cycle industry, nuclear (and later (in 1950s) thermonuclear) weapons, nuclear submarines and above water nuclear powered ships were built in this period. The scale of military programs and very high rates of their implementation, as well as a number of technological challenges to be solved, formed a basis for the problematic situation generation. Such kind of problems are currently called "nuclear legacy".

Following military tasks, challenges of nuclear power industry began to find their solutions. In 1954, firstly in the world, a nuclear power plant was put into operation in the Soviet city Obninsk. In 1964, Novovoronezh NPP began its operation, equipped with LWR-210 reactor and Beloyarsk NPP with AMB-100 reactor; in 9 years, in 1973 – Leningrad NPP with RMBK-1000. At the same time, nuclear power usage has been intensively developed in the national economy – medicine, geology, transport (building of vehicles with nuclear power installation (NPI)), etc. In some cases, introduction of new technologies was carried out without elaboration of the design solutions relating to the whole operation life of the facility. As a result, many installations used in research and industrial fields are currently elements of so-called "nuclear legacy".

Nuclear Safety Institute, IBRAE RAN, Moscow, Russia

Nuclear and radiation safety assurance was often recognized as a necessary condition for application and development of nuclear technologies. Nevertheless, a "delayed solutions" practice was often implemented. Accumulation of challenges lasted during decades, and the majority problems occurred in the course of both SNF and RW management, and decommissioning of nuclear and radiation hazardous facilities. It should be noted that a system of planned economy actual in the USSR at that time, permitted, in principle, to formulate such problems and to solve the accumulated ones, but this has not been done in time. The first grave revision of the situation in nuclear and radiation safety took place only after the Chernobyl accident. A number of nuclear legacy tasks remained outstanding due to the occurred economic crisis and following USSR break-up.

Economic problems specific to Russian Federation in 1990s, did not allow systematic solution of the accumulated problems either. Many nuclear legacy facilities still kept some resource of containment barriers at that time, and nuclear industrial enterprises could provide adequate safety using their own resources.

As a result, the ambivalent situation had been formed up to 2005–2006. On one hand, an evident progress was reached in nuclear and radiation safety of Russian Federation. On the other hand, many delayed solutions caused local (but considerable) increasing potential environmental and public risks, as well as significant complication of the operator's activity, responsible for safety assurance.

The summary of positive moments is as follows:

- The state system of safety control and regulation during nuclear energy use has already been arranged in the middle of 1990s.
- The government policy fundamentals in nuclear and radiation safety area were specified in 2003 [2].
- The environmental policy fundamentals during nuclear energy use were postulated in 2004 [3].

Recently, the majority nuclear and industrial enterprises increased their manufacture safety level considerably. The number of automated shutdown due to critical state is one of the most hazard violations in the course of operation; this number is an evidence of safe NPP unit operation. During last 15 years, this index decreased steadily, and since 1997, it is stably less than 1 (per 700 hour unit operation) [4]. Accidents and incidents occurred at Rosatom's nuclear and radiation hazardous facilities over the period 1990–2006 did not cause any significant release of radioactive substances (Table 1).

At normal operation of nuclear energy use facilities, the contribution into average public doses, due to radiation exposure, is much less than 1% (0.14%), while contribution of man-made background, due to nuclear weapon tests and radiological accidents is 0.69%. These are much less than contribution of medical exposure (29.4%) and natural radiation background (69.8%).

It is obvious that such ambivalence could not continue for a long time. The problem statement of nuclear power industry development and the approval of the relevant federal target program (FTP) [5] by the RF Government were accompanied with a request for elaboration of new federal target program directed to the

Table 1 Accidents and incidents at nuclear and radiation hazardous facilities of Rosatom (1990–2006)

Date	Facility	Event description	Number of victims	Consequences
1993, April	SCC, Seversk of Tomsk Region	Damage of process equipment with radioactivity release	None	Contamination of a part of the site, HPZ and SA without overexposure of personnel and population
1993, August	NIIAR, Dimitrovgrad of Ulyanovsk Region	Operations with irradiated target at a reactor loop	One	Radiation injury, amputation of fingers
1995, May	Smolensk NPP	Installation of gamma-radiation source dropped out of flaw detector	One	Radiation burn of fingers
1997, May	NCCP Novosibirsk	Criticality in process tank	None	No consequences
1997, June	VNIIEF Sarov	Criticality at critical assembly	One	Overexposure with fatality
1999,June	SCC, Seversk of Tomsk Region	Ejection of irradiated blocks from reactor channel	None	Irradiation of two workers, no medical consequences
2000, September	PA Mayak, Chelyabinsk Region; Beloyarsk NPP, Sverdlovsk Region	Disruption of South Urals power grid and reactor scrams	None	No consequences
2003, September	SevRAO Gremikha village, Murmansk region	Mitigation of abnormal place of SRW storage	None	Excess annual dose limits of four workers without any consequences

nuclear legacy problem solving. At that, one recognizes that, reiteration of approaches applied within the previous similar program is not productive. It should be mentioned that FTP "Nuclear and radiation safety of Russia for 2000–2006" had very restricted financing (943.2 million rubles over the whole period) [6]. Within this program, the situation was only assessed and analyzed with respect to nuclear and radiation safety assurance; and the results had been reflected in "The Fundamentals of the Governmental Policy". The systems had been developed necessary for assurance and control of nuclear and radiation safety, such as the system of the state account and control of radioactive substances and radioactive waste, The Uniformed state automated system of radiation situation control in Russian Federation, etc. Generally, the program implementation permitted to reach progress in solving common scientific and engineering problems in the field of nuclear and radiation safety assurance, but did not change situation in this area, because efficient instrumentation and appropriate conditions for effective solving these problems in future have not been developed.

It should be noted that active measures directed to nuclear legacy mitigation in 1998–2006, dealt mainly with the overall dismantlement of nuclear submarines. This fact has a simple explanation: the necessity to maintain safe conditions of facilities under overall dismantlement was very high, while ability for the responsibility transfer to the operator was practically absent.

With the purpose of a new federal target program development, interdepartmental working group had been organized under general coordination of the Federal Atomic Energy Agency. Rosatom's specialists elaborated the Program jointly with colleagues from other research institutes and departments. Rosatom is the state customer-coordinator; the state customer are: Rosatom (including facilities of RAN, RASKhN etc.), Rosnauka, FMBA of Russia, Rosstroj, Rosmorrechflot, Rosprom, Rosobrazovaniye, Rostechnadzor, Roshydromet, Emercom of Russia. IBRAE RAN was direct designer of the Program. Many chiefs and specialists participated in the program elaboration. We must mention the decisive contribution of Rosatom's Chief Mr. S.V. Kirienko, as well as Mr. A.B. Malyshev, A.M. Agapov, E.G. Kudryavtsev, and L.A. Bolshov into specification of the main program parameters. The program conception and structure of actions had been developed in the course of collective work.

The program conception [7] mentioned that the up-to-date situation in the field of nuclear and radiation safety assurance is specified by the following key factors:

- The presence of nuclear and radiation hazardous military facilities, which do not comply with the contemporary nuclear and radiation safety requirements («nuclear legacy»)
- Recognition both of accumulated problems solving necessity at the governmental level and their further postponement inadmissibility

The scale of challenges accumulated in Russian Federation is specified by the following circumstances:

- Nuclear and radiation hazardous facilities of the Federal Atomic Energy Agency (4 NPP units, 10 industrial graphite-uranium reactors and more than 110 nuclear and radiation hazardous facilities of another destination), of the Federal Agency of Industry, Federal Agency of Sea and River Transport, and other federal authorities (up to 50 facilities) had been stopped, but not decommissioned.
- Reliable environmental isolation had not been ensured at some shallow facilities for radioactive waste store; their putting into safe conditions is necessary, in addition to building of new facilities for radioactive waste disposal.

- Large amounts of radioactive wastes are not isolated from the environment (Techa reservoir cascade, storing pools and tailing dumps of nuclear fuel cycle enterprises).
- More than 18,500 t of spent nuclear fuel had been accumulated. Used storage capacity of SNF repositories are close to critical values at NPPs equipped with reactors RMBK and EGP-6 types, and radioactive waste storage facilities at these plants.
- More than 15,900 enterprises apply sources of ionizing radiation, therefore their vulnerability with respect to terrorist threat increases considerably.
- Remediation challenges did not find their regulative, legal or engineering solutions, for sites generated using nuclear explosive technologies (facilities built using peaceful nuclear explosions).
- Engineering systems of some nuclear and radiation hazardous facilities are more than 50–60 years old; they need urgent modernization.

Thus, the Federal target program "Nuclear and radiation safety assurance for 2008 and for the period till 2015" must postulate establishment of all necessary conditions, at which nuclear and radiation safety would be assured for a long perspective.

Detailed characteristics of accumulated SNF and RW are well-known: they are given in the first RF National Report «About fulfillment of engagements resulting from the Joint Convention on safe management of spent nuclear fuel and safe management of radioactive waste» [1].

When elaborating the program, it was important to show real criticality of the situation in the field of RW and SNF management as well as in other spheres. As for RW management, this criticality consists of the fact that RW treatment does not keep pace with generation of new waste, so accumulated amount of waste increases, and, up today, it is 477 million cubic meters liquid RW and 77 million tons solid RW.

In addition to demonstration of the problem solving necessity, we had to justify the selection of ways and approaches to their solution. The objective of FTP «Nuclear and radiation safety assurance for 2008 and for the period till 2015» is comprehensive solution the problem of nuclear and radiation safety assurance in Russian Federation related to:

- SNF and RW management
- Decommissioning of nuclear and radiation hazardous facilities
- Improvement of the state systems of nuclear and radiation safety assurance and control
- Public and environmental radiation safety assurance

Several options have been considered in the course of the Program elaboration; they assume implementation of different strategies: delayed solutions, guaranteeing development and intensive solution of problems accumulated. The Federal target program "Nuclear and radiation safety assurance for 2008 and for the period till

2015" approved by the RF Government is based on the development assurance strategy option. Its objective is to create all necessary conditions, under which long-term nuclear and radiation safety will be assured. The Program is aimed at putting the nuclear legacy facilities into safe conditions and at provision of the long-term governmental guarantees of nuclear and radiation safety.

Total amount of the Program financing is 145.3 milliard rubles; the major quota (131.8 milliard rubles) is due to the federal budget, 1.3 milliard rubles – to the RF subjects' budgets, 12.2 milliard rubles is a quota of off-budget funds. The federal budget funds are distributed as follows: 87.9 milliard rubles are (capital) investments; 10.9 milliard rubles – expenses for R&D; 33.0 milliard rubles – other expenses. And the amounts of financing are assumed to be increased annually.

The Program accomplishment means activity in some issues.

Within the first direction – "Building of major infrastructure elements for spent nuclear fuel and radioactive waste management", – the following actions are planned:

- Performance of some R&D works directed to development and accompaniment of the system project of the state system arrangement for SNF and RW management, involving coordination and approval of requirements for SNF and RW, storage (both temporary and longer-term) and disposal facilities, vehicles, allocation maps and operational procedures of facilities for spent nuclear fuel and radioactive waste temporary store
- Building of large facilities (for 38,000 t SNF) for SNF store from RBMK-1000 and LWR-1000 reactors, including construction of dry storage facility and reconstruction of wet storage facility at FSUE "MCC" in Krasnoyarsk Territory
- Performance of operations dealing with building of experimental industrial facility of high-level waste final isolation in geological formations
- Arrangement of experimental demonstration centre of SNF treatment using innovation technologies at the federal state unitary enterprise "Mining chemical combine"
- Construction, reconstruction and enlargement of capacities for radioactive waste management at "Radon" special combines
- Identification of the structure and building of the system of regional facilities for radioactive waste shallow disposal

The second direction – "Practical solution of legacy problems" – includes:

- Performance of comprehensive engineering and radiological examination of the legacy nuclear and radiation hazardous facilities, with respect to which there are no design solutions, in order to transfer them into safe conditions
- Implementation of decommissioning, dismantlement, liquidation and/or transfer into safe conditions of stopped nuclear and radiation hazardous facilities
- SNF removal resulted from the research installation operation, with the purpose of its further processing
- Remediation of radioactively contaminated sites, buildings and constructions

- Decommissioning and dismantlement of spent nuclear installations and sources of ionizing radiation

The third direction – "Arrangement and improvement of systems needed for nuclear and radiation safety assurance and control under conditions of routine operation and during emergencies" – includes arrangement and improvement of:

- Unified state automated system of radiation situation monitoring within the RF territory
- The state account and control systems with respect to nuclear materials, radioactive substances and radioactive waste
- Automated system of permanent monitoring of nuclear and radiation hazardous subjects (consignments) and materials, including that being performed during their transportation by all types of vehicles
- Unified state system of emergency prevention and mitigation to avoid emergencies with radiological consequences, including material and technical basis of special teams involved in radiological emergency mitigation and ensuring their preparedness

The fourth direction – "Improvement of protectability of workers, the public and the environment against radiation exposure" – includes:

- Improvement of medical care system of the personnel and performance of radiation epidemiological examinations
- Elaboration of up-to-date methodologies, development of medical treaties and equipment for prophylaxis, diagnostics, treatment and rehabilitation of the personnel from nuclear and radiation hazardous manufactures and the public damaged due to exposure to different types of radiations
- Reconstruction of medical institutions, having in mind medical care for patients with radiation injures
- Increasing radiation safety level in the course of mining, processing and use of mineral with excess contents of natural radionuclides

The fifth direction – "Scientific, information analytical and organizational provision of nuclear and radiation safety" – includes:

- Implementation of operations directed to elaboration of methods and analysis tools, as well as safety justification of nuclear and radiation hazardous facilities and their individual and environmental impact
- Activity monitoring in the field of nuclear and radiation safety assurance, generation of science-based long-term predictions, in terms of perspectives of the state social economic development, informational analytical and expert support for the Program implementation control
- Informational activity with respect to the Program implementation and the state of nuclear and radiation safety in Russian Federation

The Program actions are specific for the particular sites and facilities, and geography of their implementation covers all regions of the branch enterprises allocation.

References

1. RF National Report "About fulfillment of engagements resulting from the Joint Convention on safe management of spent nuclear fuel and safe management of radioactive waste". Moscow, 2004.
2. The RF state policy fundamentals in nuclear and radiation safety for the period till 2010 and for the further perspective (approved by the RF President on 4 December, 2003. Order-2196).
3. Fundamentals of RF Minatom's ecological policy (approved by RF Minister for atomic energy No 67 of 19.02.2003).
4. Safety report. M.: «Comtechprint» edition, 2005.
5. RF Government Directive N 605 of 6 October, 2006 "About the federal target program "Development of nuclear power complex of Russia for 2007–2010", and for the perspective till 2015" // RF Legislative Collection, 16 October 2006. N 42. Article 4380.
6. RF Government Directive N 149 of 22 February 2000 "About the federal target program "Nuclear and radiation safety of Russia" for 2000–2006" RF Legislative Collection, 28 February 2000, N 9, Article 1037.
7. RF Government Decree N 484-r of 19 April 2007 "About the Conception of the federal target program "Nuclear and radiation safety assurance for 2008", and for the period till 2015" // RF Legislative Collection, 30 April, 2007, N 18. Article 2248.

Strategy for Russian-Norwegian Regulatory Cooperation

M.K. Sneve

Abstract This paper reports on the progress within that programme in relation to remediation of the Sites for Temporary Storage (STS) operated by SevRAO at Andreeva Bay and Gremikha, as well as in recovery and decommissioning of Radio-Thermal Generators from remote locations in the Russian north. Based on this progress, the paper considers the evolution of strategy for future regulatory cooperation between Russian and Norwegian regulatory authorities and the role of wider international involvement.

Keywords Regulatory cooperation, STS, remediation work

1 Introduction

Norway has been involved in nuclear safety projects in north-western Russia since 1995 through the Norwegian Plan of Action. Cooperation between responsible authorities, identification of protection objectives, clarification of risks and recommendations to the regulatory process for nuclear projects within the Russian Federation were reported and discussed at a previous NATO workshop in 2004, on the subject of radiation and environmental safety in north-west Russia and related use of impact assessment and risk estimation.

2 Strategic Objectives of Regulatory Cooperation

Russian and Norwegian regulatory cooperation recognises that there are regional and global dimensions to the activities. The large radioactive source terms in northwest Russia present a hazard to human health and the environment in a vulnerable region.

Senior Adviser
Norwegian Radiation Protection Authority, Norway

There is a need to manage risks to workers as well to protect the public from accidents and further spread of contamination into areas adjacent to the STSs during remediation work. At the same time, there is a global interest in the prevention of proliferation of nuclear materials and technology, and control of radiation sources and radioactive waste. International cooperation in managing the global nuclear legacy is very significant, and harmonised approaches to nuclear and radiation safety supervision can be effective in consistent risk reduction.

The range of nuclear and radiation safety issues of interest is also multi-dimensional, involving different types of radioactive material and different technical challenges which have to be addressed.

The Norwegian Radiation Protection Authority (NRPA) is the national regulatory authority in Norway for radiation and nuclear safety. Its natural role in the Norwegian Plan of Action is to work with and support other national regulatory authorities. Russia is a much bigger country with a very much bigger nuclear and radiation infrastructure. As such, there is a need for NRPA to work with the separate Russian regulatory authorities dealing with safety, environmental protection and human health protection, in the military and civil sectors, and at the federal and regional levels.

The complexity of the situation and the need for improved coordination was recognised in the 2004 NATO workshop. It is pleasing therefore to be able to report the enhanced and growing relationship between NRPA, the Federal Medical Biological Agency (FMBA), the Nuclear, Industrial and Technological Regulatory Authority (Rostechnadzor) and so-called Military GAN, as well as their technical support organisations such as the Institute of Biophysics (IBPh), the Scientific Engineering Center of Nuclear and Radiation Safety (SEC-NRS), and the Inter-branch Expert and Certification Center for Nuclear and Radiation Safety (REScenter).

At the same time there is a need for regulators not to work in isolation. While each organisation must have and respond to its own functional responsibilities, efficient regulatory supervision has to involve good communications between operators, their contractors and the regulators. Again, it is good to be able to report good progress, illustrated by the participation in this workshop.

Noting the global dimension of the problem, it is also good to be able to acknowledge the continuing assistance of international and other national organisations which support the regulatory cooperation with information on the application of international recommendations, other national regulatory experience and peer review and other contributions to the cooperation outputs. We can say that the so-called 2+2 approach, involving operators and regulators from the Russian Federation and from the west, is working well.

The overall strategy is to generate confidence that investments in nuclear legacy management are being spent safely and following an effective and independent regulatory process. In the short term the objectives include efficient regulatory supervision:

- Based in Russian Federation law
- International guidance, e.g. from ICRP and IAEA and
- Other national good practice, in USA, UK and France, i.e., countries with similar nuclear legacy issues

This is intended to result in comprehensive handling of the industry projects in compliance with current Russian regulations and licensing procedures. In the longer term, the cooperation is intended to result in development of an enhanced safety culture in all operations.

Following the recommendations from the previous workshop, the Russian-Norwegian regulatory cooperation has focused on identification of the major nuclear and radiation threats and then identification of the specific needs to enhance the regulatory basis to address those threats. These enhancements may include the need for new norms and rules or guidance documents, but they can also include the design and implementation of timely procedures for application to specific tasks within the legacy management programme. The scope of cooperation therefore includes the following items which follow a logical sequence of steps.

1. Identification of major radiation and nuclear safety threats and the corresponding regulatory responsibilities.
2. Identification of need for new or modified regulations and regulatory guidance in order to meet higher level Russian Federation regulatory objectives and up to date protection objectives set out in evolving international recommendations. These include the requirements for safety, environmental and human health protection, and the content of corresponding risk assessments, And this work has to cover routine situations, emergency situations and strategic aspects of spent fuel and radioactive waste management.
3. Development of the required enhanced regulatory documents, using the 2 + 2 approach involving Russian and western expertise.
4. Development and support in implementation of an enhanced regulatory process, so that industrial projects can be efficiently supervised.
5. Safety case and license application review, according to the defined enhanced regulatory process, again with western expert involvement.
6. Monitor compliance with license conditions, according to the defined enhanced regulatory process and recognized good practice.

The following illustrates some important progress in this sequence which has taken place since the 2004 NATO workshop.

3 Regulatory Development for Decommissioning of Radio-Thermal Generators (RTGs)

As reported in 2004, Rostechnadzor had recognised a need for upgrading the regulatory framework for the safe decommissioning and disposal of the RTGs. This judgement took account of the large number of RTGs, the high hazard associated with the individual sources, and the upcoming work on their decommissioning and disposal. Another factor, given this was a new activity, was the relative lack of regulatory experience in this area.

Accordingly, a regulatory support project was set up between NRPA and Rostechnadzor and their technical support organisation Inter-branch Expert and Certification Center for Nuclear and Radiation Safety (REScenter). Additional input was incorporated provided by experts from France, the UK and Sweden. As a result of the project an official regulatory guidance document was prepared setting out "Requirements for planning and preparedness to mitigate consequences of radiation accidents occurred in transportation of nuclear materials and radioactive substances". This was registered in the Russian Justice Ministry as NP- 074- 06 and entered into force on 1 March 2007.

4 Regulatory Support in Rehabilitation of STS Andreeva Bay and Gremikha

In 2004, several regulatory projects were just commencing, addressing protection of workers, protection of the public, and medical emergency preparedness and response. The Russian authority with responsibility for radiation supervision at the STSs is the FMBA. Following the preparation and publication of an "Initial Threat Assessment of Radiological Risks Associated with SevRAO Facilities Falling within the Regulatory Supervision Responsibilities of FMBA". StrålevernRapport 2005:17. The output from this report was vital in determining the details of project plans to address priority regulatory development issues that corresponded with the plans for remediation projects at the STSs.

Regulatory documents produced within the projects have covered the topics:

- "Requirements for performance of personal radiation monitoring for personnel of FSUE SevRAO Branch No. 1"
- "Regulation for performance of radiation monitoring at FSUE SevRAO Branch No1"
- "Special features in application of ALARA principle in the work on SNF and RAW management at FSUE SevRAO Branch No1"
- "Radioecological monitoring on-site and in surveillance area in the course of conversion activities at STS of SevRAO"
- "Substantiation of organizational emergency response duties"
- "Hygienic requirements for personnel and public radiation safety guaranteeing at the stage of designing the work with SNF and RAW at FSUE SevRAO Branch No1 (R-GTP SevRAO-07)"
- "Development of criteria and norms on remediation of facilities and territories of STS at Andreeva Bay and Gremikha"
- "Substantiation of organizational emergency response duties"

The authors of these guidance documents have been able to take account of information exchange visits to operators and regulators in the USA and the UK with financial support from NATO.

In addition, a multi-agency medical response emergency training exercise was carried out at SevRAO enterprise in Andreeva Bay.

5 Conclusions and Next Steps

Regional safety and global security concerns arise concerning the nuclear legacy in northwest Russia and international cooperation is playing a big role in reducing risks. Regulatory supervision is an important part of that process and real progress has been made since the previous NATO workshop in 2004. Good working relationships have been developed and there is a good understanding of the main threats and regulatory responsibilities. The Norwegian-Russian cooperation has a proven capability to deliver significant new regulatory documentation that corresponds to the need for supervision of industrial projects.

A generic logical sequence of steps in regulatory cooperation has been developed and usefully exploited at least so far as development of regulatory requirements is concerned. It is now necessary to maintain momentum in:

- Supporting review of safety cases and licence applications
- Maintaining vigorous supervision of safety while keeping the process efficient
- Monitoring of project implementation in compliance with design requirements and licence conditions and
- Continuing improvements in coordination of emergency preparedness and response

In carrying such work forward, it is important to maintain a coherent relationship with the wider strategic goals of nuclear legacy management in the region and globally. Some common questions arise common to all legacy management programmes in all countries.

- How do you get the risk management balance right in multi-attribute problems, especially when different issues are supervised by different authorities and when the solution goes beyond their collective responsibility?
- How do you translate policy objectives into a practical regulatory supervision process?
- Does wider stakeholder engagement improve the situation through obtaining wider consent and better interpretation of policy, or hinder safety by delaying projects and confusing and diluting responsibilities?
- How do you know when you have found an adequate solution? If so, what are the tests for sufficiency?
- Or are we looking for the optimum solution? As raised in 2004, when is the best the enemy of the better?

While these are difficult questions, one clear positive response is that the answers are likely to be realised sooner through continuing coordination and international support.

Session II
Regulatory Implementation of Treaties, Standards and Recommendations

Chairmen: R. Charafoutdinov, SEC NRS and D. Louvat, IAEA

Scientific Support for Cooperation Between Regulators and Operator (2006 and First Half of 2007)

B.G. Gordon

Abstract International conventions, federal laws, federal norms and regulations serve as a legal ground of cooperation between the state regulatory body in the field of safety at nuclear energy use (regulator) and operating organizations (operator). Design and research organizations of Rosatom together with the scientific and engineering centre for nuclear and radiation safety (SEC NRS) of Rostechnadzor carries out the scientific support for this cooperation. This paper describes the interactions between these different organisations in nuclear legacy management.

Keywords Rostechnadzor, regulator-operator cooperation

The main characteristics of cooperation between the regulator and operator result from three fundamental principles:

- Division of responsibility during nuclear energy use
- The regulatory body independence upon bodies and organizations using nuclear energy
- Full responsibility of the operator for the facility safety

Main categories and amount of facilities involved in peaceful use of nuclear energy under the regulator's supervision are as following:

1 Nuclear Power Plants

- 10 nuclear power plants (31 units in operation, 4 units are being prepared for decommissioning, 3 units are under construction)

Director of the Scientific and Engineering Centre for Nuclear and Radiation Safety (SEC NRS), Professor of Moscow Engineering Physical Institute, Russia

2 Nuclear Fuel Cycle

- 15 commercial reactors (5 in operation, 10 under decommissioning)
- 27 nuclear installations on nuclear materials reprocessing
- 12 storage facilities for nuclear materials, spent nuclear fuel and radioactive waste
- 3 underground disposals for liquid radioactive waste
- 3 storage facilities for radioactive waste under decommissioning

3 Research Nuclear Installations

- 79 research nuclear installations (67 in operation, 9 under decommissioning, and 3 under construction)

4 Radiation Hazardous Units of National Economy

- 2,162 organizations, enterprises and institutions (6,397 units) – enterprises of aircraft, metallurgy, shipbuilding and chemical industries, mining and ore mining and processing branches, fuel and energy complex enterprises, geological and scientific organizations, military units and organizations of armed forces of the Russian Federation, medical institutions, custom authorities, etc.

5 Atomic Fleet

- JSC "Murmansk shipping company" – nine nuclear-powered ships (eight in operation) and six nuclear maintenance vessels (NMV) (four in operation, one under decommissioning, one decommissioned)
- FSUE "Atomflot" – nuclear-powered ships and NMV stationing, NPI equipment repair, storage and reprocessing of radioactive waste, conveying-loading activities and technological operations with nuclear fuel

6 Shipbuilding Facilities and Shipyards of Rosprom

- JSC "Baltiyskyi zavod (plant)", Shipyard "Nerpa", FSUE "PO "Sevmash", FSUE "GMP "Zvezdochka", JSC "Amurskyj shipbuilding facility", "Vostok", FSUE "DVZ "Zvezda"

7 FSUE "NITI After A.P. Aleksandrov" of Rosatom

- Benches – prototypes of ship NPF (two benches in operation, two benches under preservation)

Each State solves the problem of the regulation effectiveness in its own way. The criteria of the regulation effectiveness in Russia are based on the safety ideology, the practice of which is postulated and discussed in [1]. Rostechnadzor permanently discusses these criteria and now they are as follows:

1. Compliance with requirements of RF legislation in the field of nuclear energy use by legal entities and citizens
2. Absence of excess of main dose limits of occupational and public exposure
3. Absence of accidents with radiological consequences at the nuclear energy use facilities
4. Absence of breaching of safe operation limits at nuclear energy use facilities
5. Absence of unauthorized releases and discharges
6. Absence of stealage, losses or unauthorized use of nuclear materials and radioactive substances
7. Absence of unauthorized intrusion at the territory of the nuclear energy use facility, unauthorized access to nuclear materials and radioactive substances;
8. Results of regulatory impacts for safety assurance of nuclear and radiation hazardous facilities of past activity (radiation legacy)

It should be noted that criteria 2, 3, 6 and 7 allow **quantitative** assessment of both the facility safety and regulation effectiveness. Figure 1 demonstrates the effectiveness assessment made by specialists from SEC NRS.

Criterion	NPP	NFC	National economy	RNI	Ships
1	+	–	+	+	+
2	+	+	+	+	+
3	+	+	+	+	+
4	+	+	+	+	+
5	+	+	+	+	+
6	+	+	+	+	+
7	+	+	+	+	+
8	–	+	+	+	+

Fig. 1 SEC NRS assessment of activity effectiveness on the base of criteria 2

Regulatory impacts, which the RF legislation considers as functions of the regulator, include:

- Inspections at the nuclear energy use facilities, sanctions for violation of regulations
- Licenses for types of activities in the field of nuclear energy use
- Permissions for workers to carry out operations in the field of nuclear energy use
- Regulatory control of safety at nuclear energy use facilities
- Safety assessment of nuclear energy use facilities

These functions are in the full compliance with the convention [2]. Figures 2–7 illustrate the particular data relating to results of regulator-operator cooperation in 2006 and in the first half of 2007.
Scientific support of regulation is performed in a number of directions:

- Implementation of branch-independent safety review of nuclear energy use facilities, including assessment of calculations, **for scientific justification of decision making on licensing or license withdrawal**
- Creation and improvement of regulatory and technical documentation, including analysis of the practical **use of this documentation in supervision and review** of safety
- Creation and input of data bases of reliability, violations and failures, using analysis of violations **to settle feedback with safety review**
- Participation in surveys of nuclear energy use facilities to collect **own information** on the conditions of safety important equipment and its life time estimation at the Service

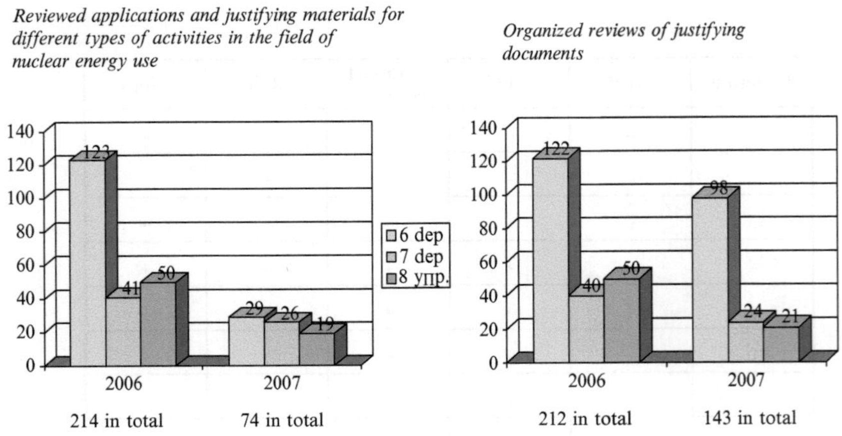

Fig. 2 Licensing and authorization activities of departments of Rostechnadzor headquarters in 2006 and in the first half of 2007

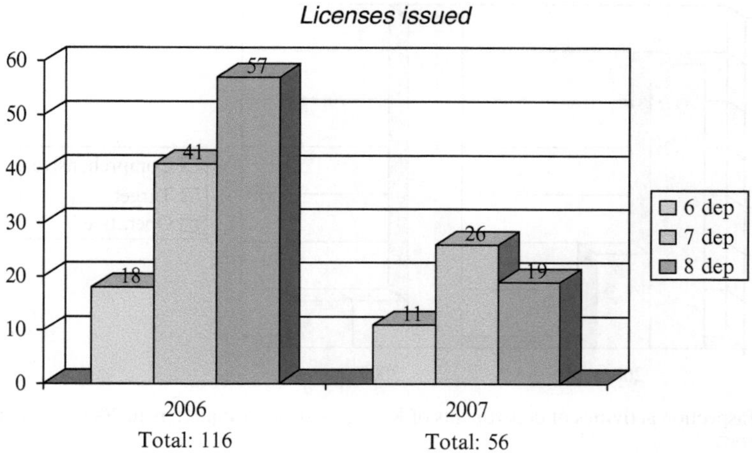

Fig. 3 Licensing and authorization activities of departments of Rostechnadzor headquarters in 2006 and in the first half of 2007

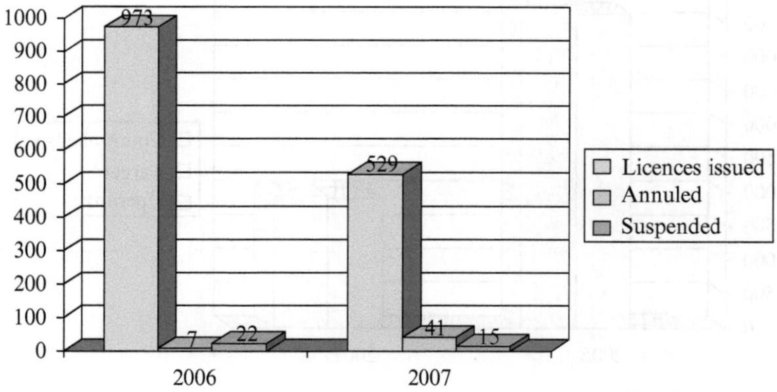

Fig. 4 Data on licensing activity of regional offices on nuclear and radiation safety supervision in 2006 and in the first half of 2007

- Organization of **training and methodic centre** for professional development of authority personnel, and in case of accidents for submission of evaluated value of accident parameters to information and analytical centre of the Service
- Organization and implementation of scientific and research work necessary for scientific justification of above-mentioned directions

The regulator defines these directions and specifies them annually in the annual plans. SEC NRS is the technical support organization, so specification of special features of scientific activity is an important issue of self-identification and self-knowledge of its objectives:

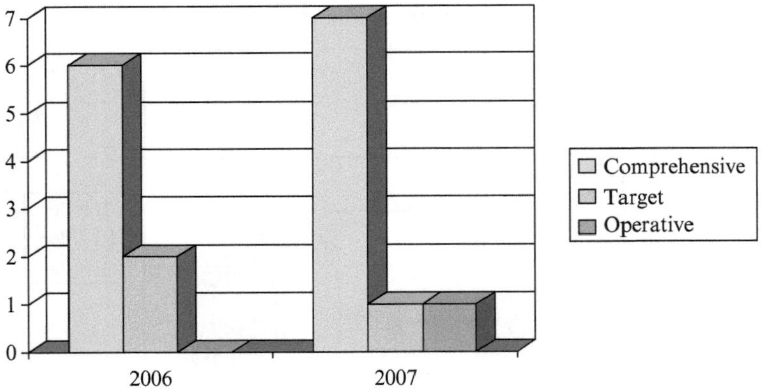

Fig. 5 Inspection activities of departments of Rostechnadzor headquarters in 2006 and in the first half of 2007

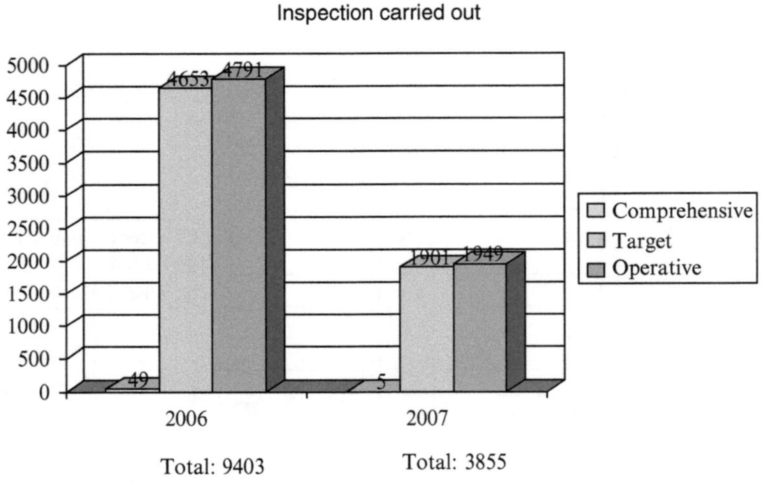

Fig. 6 Data on supervision activities of regional offices on nuclear and radiation safety supervision in 2006 and the first half of 2007

- Control, analysis and assessment of **reliability and quality of scientific and research work** performed by the industry for safety justification and to be submitted to Rostechnadzor for obtaining a license
- Development of proposals on **scientific and research work performance, which are included in the license conditions issued by Rostechnadzor**
- Participation in scientific and research work for development and **justification of criteria and principles of nuclear and radiation safety which are used** in reviews and regulatory documents

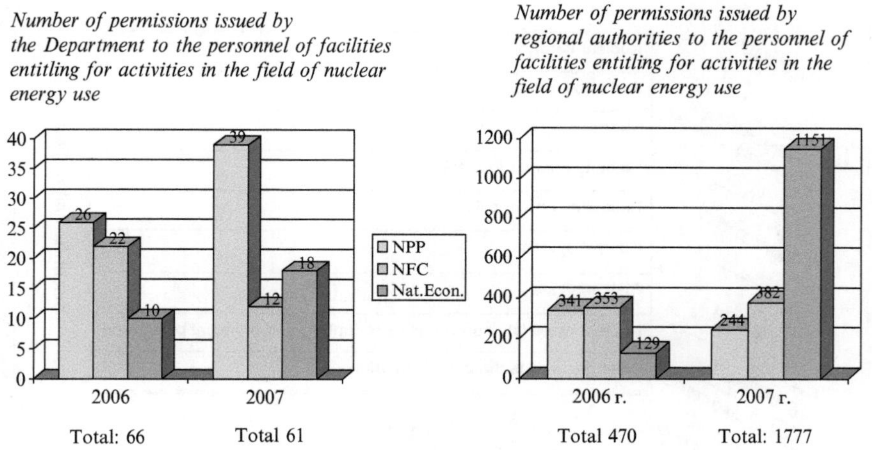

Fig. 7 Number of permissions issued to personnel of nuclear energy use facilities entitling for activities in the field of nuclear energy use in 2006 and in the first half of 2007

- Concentration of scientific and research work at first levels of defence-in-depth
- **Summarizing of experience of reviews and other scientific and research works performance in the sphere of nuclear and radiation safety (development of review guidelines, methodical documents and safety guides)**

Due to these features, SEC NRS activity differs drastically from Rosatom's organizations, which implement scientific support of the operator. With the purpose of successful implementation of its functions, some special structure of SEC NRS departments is arranged; Fig. 8 shows this structure.

Such approach makes special requirements for technical provision of SEC NRS specialists. The main tools include:

- Analytical simulators for operative modeling of NPP accidental parameters
- Office typography
- Electron microscope with necessary equipment
- Laboratory of spectrometric measurement with portable detectors
- Laboratory of estimation of radiochemical processes explosibility
- Work places on the base of personal computers

Information provision is rather important:

- Data base on regulatory and technical documents
- Data base on failures at nuclear energy use facilities
- Data base on several hundreds experts of the third party organizations engaged by SEC NRS
- Availability of several thousands expert reports on actual problems of almost all nuclear energy use facilities
- Data base on certified software

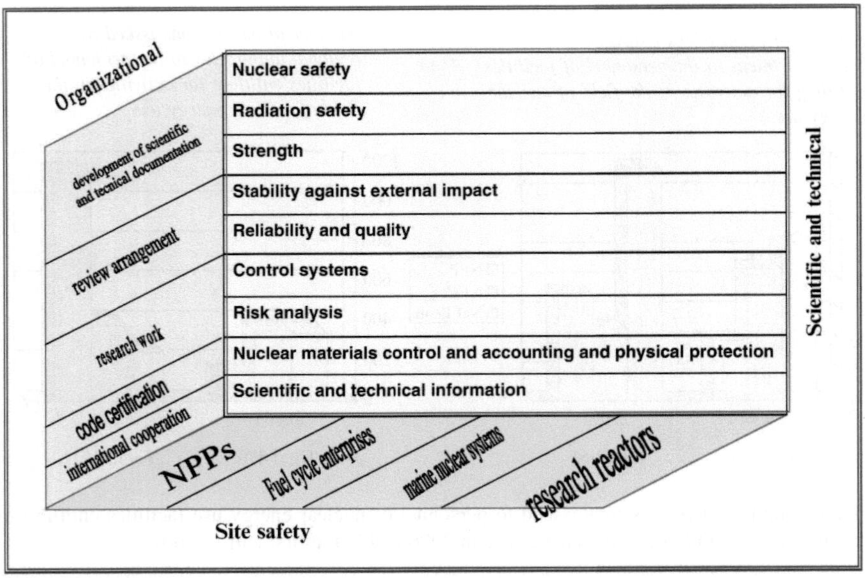

Fig. 8 SEC NRS departments

- Maintenance book of preliminarily calculated NPP emergency states
- Library

Figures 9–12 illustrate main results of the regulator's scientific support and dynamics of SEC NRS active staff.

Generally, scientific support for cooperation between regulators and operators may be declared as satisfactory. At the same time, contemporary development of nuclear power industry in Russia and increasing Rostechnadzor's requirements for a quality of scientific support pose some relevant problems for SEC NRS:

- Engagement of young specialists
- Assurance of succession and knowledge transfer
- Young specialists probation at nuclear energy use facilities
- Engagement of inspectors to SEC NRS activities
- Federal Law development "On State Regulation of Nuclear and Radiation Safety"
- Development of the recommended terms glossary
- Upgrading and harmonization of FRR system
- Development of technical regulations
- Mastering of procedures of peer review and projects expert maintenance
- Implementation of RAIS into regulatory practice
- Participation in the Federal Target Program of Rosatom and Ministry of Emergencies
- Completion of building reconstruction in 2007

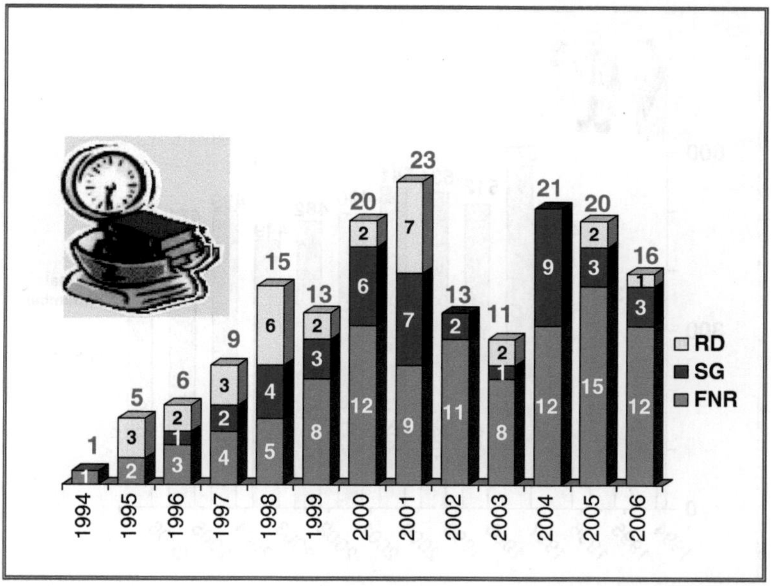

Fig. 9 Regulatory documents development dynamics

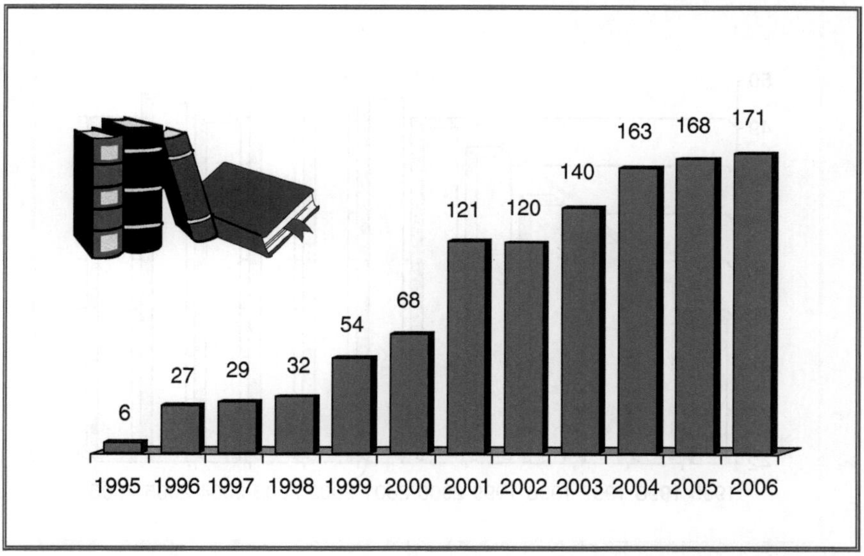

Fig. 10 Dynamics of review reports development

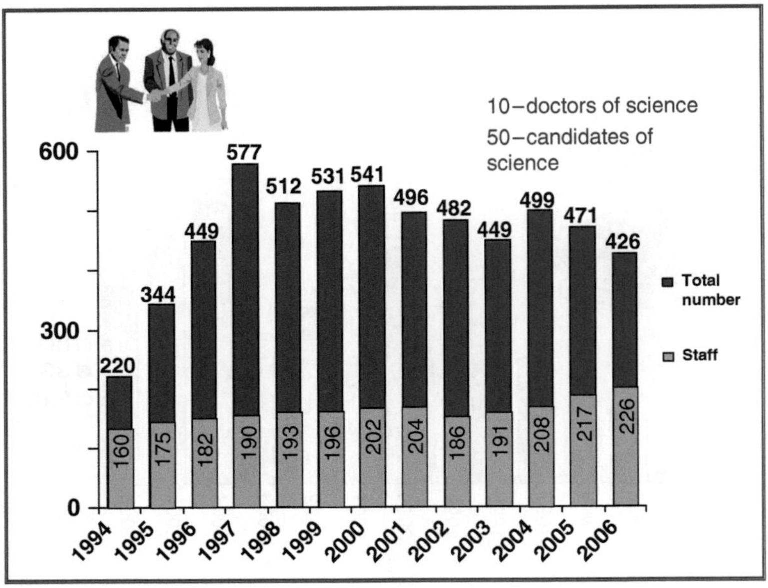

Fig. 11 Dynamics of SEC NRS quantity

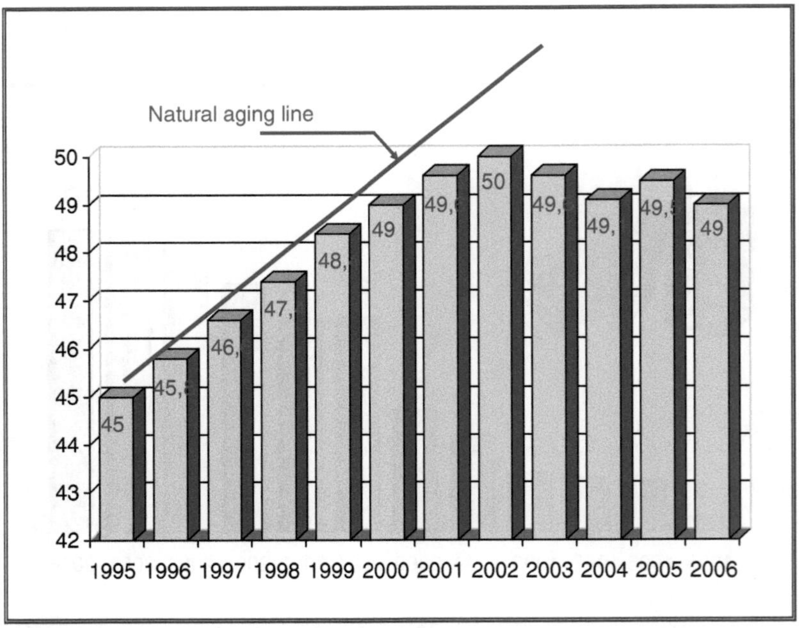

Fig. 12 Average age of SEC NRS employees

While the regulatory body faces with problems of its activity improvement:

1. **Upgrading of regulatory legal base**

   - Legislation in the field of nuclear energy use
   - Technical regulation
   - Federal rules and regulations
   - Guidelines of the Service

   1.1. Development of the concept of improvement of federal rules and regulations
   1.2. Development of the draft law "On State Regulation of Nuclear and Radiation Safety"
   1.3. Analysis of regulatory documents in force and their actualization

2. **Licensing and authorization activity**

   2.1. Drawbacks of social support to inspector staff, high average age and understaffed inspection divisions of the regional offices
   2.2. Lack of system of centralized training and professional development of regional offices inspector staff
   2.3. Methodical elaboration of licensing and authorization activity with the aim of concentration of resources at licensing of the types of activities of nuclear and radiation hazardous facilities
   2.4. Accurate elaboration (legal and technical) of issued license conditions

3. **Development of organizational arrangements for creation of system of comprehensive training of regional office inspector staff**

In conclusion, we can recognize that implementation of cooperation between regulators and operators in Russia is rather dynamic and complies with both national legislation, and international practice.

References

B.G. Gordon, 2006, Safety ideology. M. SEC NRS.
M. Znanie, 2003, Convention on nuclear safety. //Safety of Russia.

Strategy for the Environmental Regulation of Remediation and Decommissioning at Sellafield

A. Mayall

Abstract The Sellafield site on the coast of Cumbria in north west England represents a major part of the UK's nuclear legacy. Sellafield, which includes the smaller Windscale site within its boundary, is a large complex nuclear chemical facility. Sellafield and Windscale have supported the civil nuclear power programme since the late 1940s and in the early days were involved in plutonium production for the UK's nuclear weapons programme.

Sellafield, which is owned by the Nuclear Decommissioning Authority and operated by Sellafield Ltd., is undergoing a transition from production operations to decommissioning and clean up. Decommissioning and clean-up will involve the currently operational facilities and the legacy facilities which contain radioactive wastes from the early days of civil and military operations. In this period of change it is important that the strategic long-term aspects of the regulation of Sellafield are managed so as to prevent and minimise future environmental impacts and risks.

To achieve the best environmental results the Environment Agency recognises that traditional direct regulatory activity such as authorisation (permitting), compliance assessment and enforcement, needs to be supplemented with partnership working and regulatory advice and influence.

The Environment Agency is the principal environmental regulator at nuclear sites in England and Wales. Its regulation at Sellafield is carried out by the Nuclear Regulation Group (North) (NRG(N)).

Keywords Sellafield, environmental principles, strategic objectives at Sellafield

1 Introduction

The Sellafield site on the coast of Cumbria in north west England represents a major part of the UK's nuclear legacy. Sellafield, which includes the smaller Windscale site within its boundary, is a large complex nuclear chemical facility.

Environment Agency of England and Wales, Team Leader (Sellafield)
Nuclear Regulation Group (North), UK

Sellafield and Windscale have supported the civil nuclear power programme since the late 1940s and in the early days were involved in plutonium production for the UK's nuclear weapons programme.

Sellafield, which is owned by the Nuclear Decommissioning Authority (NDA) and operated by Sellafield Ltd. (SL), is undergoing a transition from production operations to decommissioning and clean up. Decommissioning and clean-up will involve the currently operational facilities and the existing legacy facilities which contain radioactive wastes from the early days of civil and military operations. In this period of change it is important that the strategic long-term aspects of the regulation of Sellafield are managed so as to prevent and minimise future environmental impacts and risks.

This paper summarises the strategic approach to our regulation at Sellafield, developed by the Environment Agency's Sellafield Team.

2 The Environment Agency's Role at Nuclear Sites

The Environment Agency has two Nuclear Regulation Groups (NRG), one covering nuclear sites in the north of England and Wales and the other covering sites in the south. NRG (North) is based at the Environment Agency's offices at Penrith in Cumbria and includes a team of Nuclear Regulators which covers Sellafield.

The NRGs work to ensure the protection of the public and the wider environment from radiation, to prevent pollution, to protect and enhance the environment and to contribute to the UK's aim of sustainable development. This is achieved through influence and advice in addition to licensing/authorisation, compliance assessment and enforcement under legislation such as the:

- Environment Act 1995 (sets out regulatory powers etc.)
- Radioactive Substances Act 1993 (RSA 93) (which deals with the disposal and discharge of radioactive waste/effluents from nuclear sites) and the
- Pollution Prevention and Control Regulations 2000 (PPC) (which covers the environmental regulation of industry and the control of non-radioactive pollution)

We collaborate with a number of other Environment Agency groups and functions to ensure an integrated approach to environmental protection. We also work with a wide range of external stakeholders – local, national and international. In particular we have close working arrangements with our colleagues in the Nuclear Installations Inspectorate (NII) of the Health and Safety Executive (HSE). We maintain good contacts with the Food Standards Agency, the Nuclear Decommissioning Authority, other regulatory bodies, local authorities and, our sponsoring department, the Department for Environment, Food and Rural Affairs. We also have good links with our regulatory counterparts abroad, for example in Norway, Ireland and France.

3 The Environment Agency's Corporate Strategy

The Environment Agency's corporate strategy, 'Creating A Better Place', sets out a vision for the environment in England and Wales of a better place for people and wildlife, for present and future generations [1]. To achieve this vision our Corporate Strategy sets out nine environmental goals (see Fig. 1 below). All of these are relevant to our regulation at Sellafield.

To enable us to deliver these goals we perform five roles as follows.

An efficient operator
We focus on environmental outcomes. We carry out our regulatory role in a way that gets the maximum benefit for the environment from public and charge payers' money.

A modern regulator
We work to prevent pollution and protect people and the environment from harm. We aim to find the right balance – a proportionate, risk-based response, that will drive environmental improvements (both short and long term), reward good performance, but still provide the ultimate reassurance that tough action will be taken on those who fail to meet acceptable standards. We aim to be:

- Transparent – we must have rules and processes which are clear to those in businesses and local communities
- Accountable – we must explain ourselves and our performance

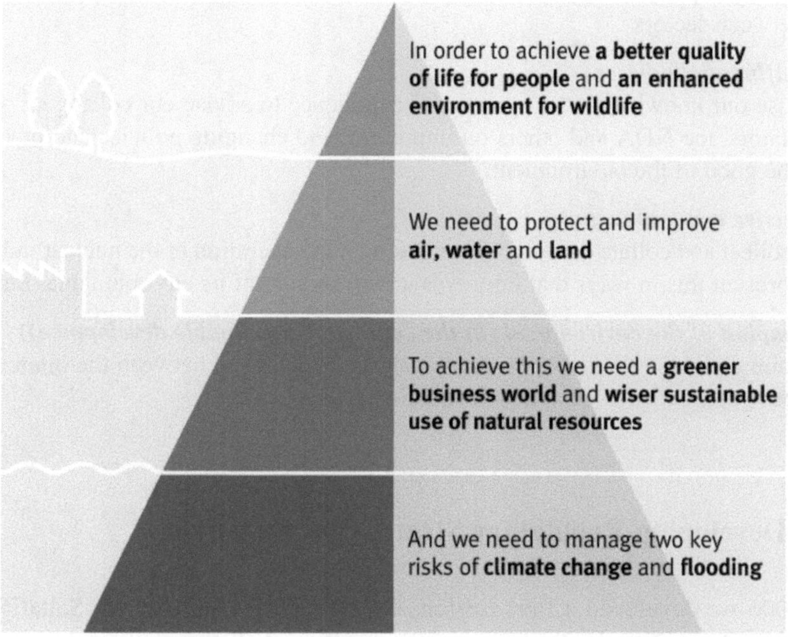

Fig. 1 The Environment Agency's environmental goals

- Consistent – we must apply the same approach within and between sectors and over time
- Proportionate (or risk-based) – we must allocate resources according to the risks involved and the scale of outcomes which can be achieved
- Targeted (or outcome-focused) – the environmental outcome must be central to our planning and in assessing our performance

Part of this approach to 'modern regulation' is to encourage businesses to keep the environment at the centre of their thinking and to identify joint objectives. To achieve these objectives, we are identifying and addressing the priority environmental issues resulting from individual industry sector activities and have developed a generic environmental improvement plan for the nuclear sector – the Nuclear Sector Plan [2].

The nuclear sector has reduced its impact on the environment, particularly its radioactive discharges to air and water, over the past 20 years. But the industry and the environment still face serious challenges, for example, the targets for environmental discharges are getting tougher, especially in relation to the marine environment.

The purpose of the Nuclear Sector Plan is to:

- Focus on the most significant risks and impacts that the sector poses to the environment
- Deliver improvements in the sector's environmental management and performance
- Prioritise and target Environment Agency effort within and across sectors
- Achieve, through co-operation with sectors, environmental benefits beyond those that can be achieved through direct regulation alone and
- Monitor progress in delivering environmental improvements, within and between sectors

An influential adviser
We use our knowledge and operational experience to advise our colleagues, other regulators, the NDA and others on improving and changing policies and practices for the good of the environment.

An active communicator
We collect and collate data and information on the operation of the nuclear industry and present this in ways that improves understanding of its environmental impact.

Champion of the environment (in the context of sustainable development)
We aim to ensure that an appropriate balance is achieved between the interests of the environment and social and economic needs.

4 Developing Regulatory Strategy for Sellafield

In 2006 we developed a first version of a Regulatory Strategy for Sellafield to provide a framework for the planning and direction of all our regulatory work at the site and to provide a link to the wider Environment Agency strategies such as the Corporate Strategy described above and our general strategy for the regulation of radioactive substances [3].

Our strategy for Sellafield includes reference to our Environmental Principles, strategic objectives and sets out what we expect of the operator of the Sellafield site. We recognise that our strategy needs to continually develop and respond to changes in Environment Agency strategy and priorities, NDA strategy, nuclear site strategy and plans, and UK policy etc.

4.1 Environmental Principles

Our strategic direction is guided by a framework of principles. Nationally, we are developing Environmental Principles (EPs) that will form a consistent and standardised framework for the technical assessments and judgements that we must make when regulating radioactive wastes and substances, and nuclear site clean up [4]. The EPs underpin our regulatory decision making, including those about permitting and compliance where we regulate directly and those where we are consultees or have influence.

Our fundamental principles relate to the following:

Sustainability – Radioactive substances shall be managed to avoid placing a burden on future generations and their environment such that it compromises their ability to meet their needs.

Stakeholders – To give confidence that the right decisions will be made for the right reasons, citizens, communities and organisations shall have access to information relating to radioactive substances, key decisions shall be informed by their views, and the right to justice shall be respected.

Integrated Planning – All radioactive substances shall be managed within integrated strategies that plan their complete lifecycle taking account of all interactions, dependencies and principles.

Selecting and Implementing Management Options – The best available option for the management of radioactive substances shall be used after systematic consideration of alternatives. Consideration shall include human health, safety, the environment, waste prevention, minimisation and disposal and other likely costs and benefits.

Protecting Human Health and the Environment – Radioactive substances shall be managed to ensure an acceptable level of protection to human health, wildlife, organisms and the wider environment, and compliance with relevant dose limits and constraints is achieved. Monitoring and assessment shall be undertaken to inform decisions about radioactive substances and to establish the state of the environment.

Regulation – Regulatory systems for radioactive substances shall be independent, seek best practice through high standards of management, take account of risk, be transparent, accountable, consistent and targeted.

Best Scientific Knowledge – Decisions on radioactive substances shall be informed by the best scientific knowledge. Appropriate research shall be undertaken to facilitate technology development, to promote innovative solutions and where significant gaps in knowledge are recognised.

Uncertainties and Precautionary Principle – Decisions about radioactive substances shall take into account uncertainties and where there are threats of serious or irreversible damage, lack of full scientific certainty shall not be used as a reason for postponing cost effective measures to prevent environmental degradation.

Polluter Pays – Producers, owners and users of radioactive substances shall be accountable for the costs of managing and disposing of their radioactive substances, for associated regulation and research and for rectifying environmental damage.

Justification/Intervention – Benefits and detriments arising from practices or interventions involving radioactive substances shall be considered to establish whether the practice or intervention is justified.

4.2 Strategic Objectives

Our strategic objectives are those things that need to be achieved in order to realise the Environment Agency's environmental goals and have been set after taking account of an analysis of the issues and challenges at Sellafield, the Environment Agency's wider strategic direction and our guiding principles (see Section 4.1). The greatest benefit comes from a proactive nuclear site operator that recognises the issues and has sound plans to address them – therefore our expectations of the operator are set out in our strategy. Table 1 sets out our strategic objectives.

5 Implementation of the Strategy

To implement our strategy we need to make sure that we have the right processes in place, that we work in partnership with others wherever necessary and that our expectations of the site operator are met. This section summarises what we have in place to ensure that our regulatory strategy becomes a reality.

5.1 Annual Operational Planning

We have an annual operational planning cycle which sets work scope, priorities and allocates resource. Our annual operational plans are aimed at delivering our short-term objectives and continue to build the platform for achieving our strategic objectives. Our plans take account of our core business activities below.

Table 1 Our strategic objectives at Sellafield

Objective	
1	We will ensure that permits and authorisations are appropriately reviewed and revised to ensure that they are up to date, flexible and fit for purpose.
2	We will continue to assess compliance with limitations and conditions of the permits and authorisations.
3	We will seek environmental benefits beyond those achieved by compliance alone.
4	We will seek the reduction of the risks to the environment from hazards at Sellafield and the avoidance of additional 'legacy' issues, which could add to future environmental impact.
5	We will ensure that the site Integrated Waste Strategy (IWS) undergoes progressive improvement.
6	We will ensure waste forms and packages are demonstrated to be disposable, minimising the potential for future rework and environmental implications for the future.
7	We will require the operator to demonstrate that their forward planning for infrastructure and asset care is consistent with minimising environmental impact both in the short and long term.
8	We will influence the development of national/international guidance on Best Available Techniques (BAT) for the nuclear industry.
9	We will influence NDA strategy so that it provides appropriate direction for waste minimisation and other environmental outcomes – setting the vision and overall aims and objectives for Sellafield.
10	We will encourage the continued application of integrated fuel management plans so that environmental impacts across the whole fuel cycle are prevented or minimised for example the Magnox Operating Plan (MOP) and an Oxide Fuel Operating Plan (OOP).
11	We will encourage the development of the national IWS to provide focus to waste issues, identify waste synergies, reducing bottlenecks and ensuring appropriate waste transfers are transparent and accord with the Best Practicable Environmental Option (BPEO).
12	We will support and encourage the development of the multi-agency nuclear emergency planning and response arrangements.
13	We will facilitate the understanding of the wider environmental benefits (local and national) that can be derived from the major investment in nuclear site clean up by the NDA and the new focus on integrated waste management and environmental restoration.
14	We will ensure appropriate management of land contaminated with radioactivity and other pollutants and the protection, and if necessary, remediation of groundwater.
15	We will continue to be pro-active with respect to research and development in those areas that are relevant to our key objectives and will act to ensure or encourage waste producers and NDA to carry out appropriate research in those areas.
16	We will monitor, and where appropriate assess, organisational change to give early indication of deterioration in environmental performance at nuclear sites as a result of changes in either the Parent Body Organisation or the authorisation holder.

5.2 *Core Business Processes*

5.2.1 Compliance Assessment

Compliance assessment covers all those activities associated with checking compliance with the limits and conditions of authorisations and permits. It includes individual and team site inspections or audits, and also the review and assessment of information, reports and data.

This process includes judging the 'fitness' of the 'environment case' and procedures used by the operator to justify operation in terms of its environmental impact. This includes the assessment of the documentation and management arrangements related to the environmental protection measures of plant, processes, operations and organisation, and the inspection of the plant and procedures against the information provided in the 'environment case'.

5.2.2 Enforcement and Incident Response

This is the process, in cases of non-compliance, of carrying out enforcement action in line with the Environment Agency's Enforcement and Prosecution Policy.

This process also includes our response to events and incidents which may extend, in the case of incidents with off-site impact, to carrying out our role as part of our commitment to a multi-agency response.

5.2.3 Authorisation (or Licensing)

This is the process of granting, revoking and varying authorisations or permits.

5.2.4 Influencing

This process includes the communication of our long-term goals and objectives to ensure they are taken into account by operators when they undertake long term strategic planning.

5.2.5 Standards and Advice

The process of developing and promulgating guidance on compliance and environmental standards and responding to requests for advice on environmental issues.

5.2.6 Research

It is important that we, the operators and others, have the scientific and technical knowledge to make judgements about the adequacy of measures or plans to treat or dispose of waste. This process relates to the identification of the research needed to improve environmental protection and radioactive waste management, its commissioning and the promulgation of its results.

5.3 Partnership Working

To realise our vision and strategy we need to apply more than the statutory regulatory framework. Influencing and working with the Sellafield site operator, NDA, NII, local authorities and other stakeholders are key to success.

6 Conclusions

The UK faces a considerable task in cleaning up the legacy of the nuclear industry. The Sellafield site represents a major part of this legacy, which must now be dealt with in the interests of sustainable development. This needs to be done through early reduction in onsite risks (to avoid leaving a burden to future generations) while causing minimal impact on the environment.

Our developing strategy sets out how we believe we can contribute to this task through both our direct regulation and influence at the site. We have started to assess the environmental challenges that this task presents and are developing objectives, for ourselves and the operator, to start to address the issues raised. We recognise that this is only the start of the journey and that we will need to regularly review and monitor our strategy and progress.

Acknowledgements The author would like to thank members of the Environment Agency's Sellafield Team (Dr. M. Emptage, Mr. P. Orr, Ms. N. Lawton and Dr. I. Lowles), who worked with the author in developing the first version of our Regulatory Strategy.

References

1. Environment Agency, Creating a Better Place, Corporate Strategy 2006–2011. http://publications.environment-agency.gov.uk/pdf/GEHO0406BKFW-e-p.pdf
2. Environment Agency, Sector Plan for the Nuclear Industry. Version 1, November 2005. http://publications.environment-agency.gov.uk/pdf/GEHO1105BJVE-e-e.pdf
3. Environment Agency, Radioactive Substances Regulation: A Strategy for 2006–2011. Issue 1, December 2006. http://publications.environment-agency.gov.uk/pdf/GEHO1206BLTQ-e-e.pdf
4. Environment Agency, Radioactive Substances Regulation Environmental Principles (Interim). Version 1, November 2005. http://publications.environment-agency.gov.uk/pdf/GEHO0606BLSO-e-e.pdf

State Supervision of Nuclear and Radiation Safety During Dismantlement of Decommissed Nuclear Powered Ships and Remediation of Former Shore Technical Bases of the Northern Fleet

S. Testov

Abstract This paper describes the role of the Department of the state supervision of nuclear and radiation safety of the Ministry of Defence in the Management of decommissioning of military installations.

Keywords Supervision, regulation of nuclear and radiation safety, licensing, hazardous operations

According to the legislation of the Russian Federation, RF Ministry of Defense (MoD) performs the state supervision of nuclear and radiation safety in the course of design, manufacturing, test, operation and dismantlement of military nuclear powered installations. The Department of the state supervision of nuclear and radiation safety of MoD (hereinafter referred to as Department) was arranged in 1996 with the purpose of successful implementation of such supervision with respect to military nuclear powered installations.

That is why, our Department performs the state supervision of nuclear and radiation safety in the course of design, operation and dismantlement of nuclear submarines and nuclear powered above water ships, their serviced vessels, units of nuclear powered ships maintenance, Rosatom's and Rosprom's enterprises and organizations, involved in activity using military nuclear powered installations, spent fuel of nuclear ships, as well as during remediation of former shore technical bases of the Northern and Pacific Fleets.

In Russian northwest, the Department's supervision covers such federal state unitary enterprises as «PA «Sevmash», «MA «Zvezdochka», «NIPTB «Onega», «Severny rejd», «SRZ «Nerpa», ship repairing plants of the Navy in Polyarny, Murmansk, Roslyakovo, bases for station of nuclear submarines and nuclear powered above water ships, as well as FSUE «SevRAO» with its branches in Andreeva bay, Gremikha and Saida bay. The mentioned enterprises will remain

Department of the state supervision of nuclear and radiation safety of the Russian Ministry of Defense, Head of Inspection, Moscow, Russia under the Department's supervision even after change of type of ownership, planning in the next year.

As is well known, according to IAEA recommendations, regulation of nuclear and radiation safety during nuclear energy use includes:

- Development of main requirements for nuclear and radiation safety; regulatory and technical documents in the field of nuclear energy use
- Development of requirements for licensing and certification of technologies, objects, products etc.
- Development of guidance documents in the field of inspection and review
- According to the same recommendations, supervision includes:
- Licensing and certification
- Implementation of inspection functions
- Performance of review

In the field of nuclear energy use out of MoD's supervision, i.e., in the field of "peaceful atom" use, the Federal service of environmental, technological and nuclear supervision (Rostechnadzor) implements functions of regulation and supervision of nuclear and radiation safety in the course of nuclear energy use.

The special feature of operation of the Department is its responsibility not for all regulatory and supervision functions, but only for some part of them:

- Development of guidance documents in the field of inspection
- Implementation of inspection functions

The Department closely cooperates with different federal bodies and authorities in all other actions relating to regulation and supervision, such as:

- Analysis of nuclear and radiation safety state, physical protection of facilities, regulatory basis in the field under supervision and elaboration of proposals on improvement of the system of safety assurance of nuclear facilities, nuclear and radioactive materials
- Participation in preparation of legislative and other regulatory and legal acts
- Coordination of the state standards, regulations and norms in the field under supervision
- Coordination of requirements for the level of the personnel qualification responsible for nuclear and radiation safety
- Approval of technical tasks, design and construction tasks, as well as engineering design of nuclear installations, designs of nuclear and radiation-hazardous facilities, etc.

Several federal authorities cooperate in the close contact in the field of regulation of nuclear and radiation safety:

- Activities in the field of nuclear and radiation is implemented in compliance with regulative documents, being developed jointly with Rosatom and Rosprom, approved by Ministry of Defense of Russia (our Department).

- Licensing of activity of subordinated enterprises is performed by Rosatom jointly with the Department, according to the RF Government Directive No. 471 of 2000.
- Certification is performed by Rosatom jointly with the Department.
- Review is carried out by working teams of representatives from different organizations, including those from the Department.
- The State supervision of nuclear and radiation safety and that of physical protection are implemented by the Department, according to the Decree of the RF President of 2004 No. 1082.

In the field of licensing of facilities under supervision, the Department is responsible for coordination of documents and information justifying assurance of nuclear and radiation safety.

The state supervision of nuclear and radiation safety of FSUE "SevRAO" activity, as well as that of other enterprises and organizations, falling in the Department's responsibility, consists of control over compliance with requirements of federal laws and federal norms in the field of nuclear and radiation safety, as well as departmental regulatory documents developed by Rosatom and Rosprom jointly with the Department of the state supervision.

In addition to the state supervision of nuclear and radiation safety, our Department is responsible for issue of permissions to carry out some potential nuclear- and radiation-hazardous operations in the course of design, operation, and decommissioning of nuclear power installations, as well as during remediation of former military sites, in the course of which in some cases, abnormal situations and accidents can occur; they can cause overexposure of workers and the public, as well as the environmental contamination exceeding maximum permissible values.

As for enterprises under supervision, our Department issues permissions to carry out the following potential hazardous operations:

- Transportation of SNF to reprocessing or storage
- Re-loading of spent fuel assemblies to new canisters
- SNF discharge from reactors of nuclear submarines and from facilities for spent nuclear fuel store (both coastal and floating)
- Radioactive waste treatment, generated in the course of operation and dismantlement of nuclear submarines
- Transfer of reactor compartments of dismantled nuclear powered ships from some facilities to others
- Acceptance of reactor compartments of dismantled nuclear powered ships by the facility
- Transfer and acceptance of radioactive waste from some facilities to others and many other types of operations

Having in mind results of supervision, the Department got the right to forbid performance of potential nuclear- and radiation-hazardous operations in case when safety actions taken do not meet requirements of legislative and regulatory documents.

According to the Provisions of the Department, approved by Rosatom, development of all projects relating to nuclear and radiation hazard must be coordinated with our Department, beginning from the stage of technical task for the design. In this light, the Department also performs supervision of compliance with requirements for nuclear and radiation safety at the stage of construction of the facility for long-term storage of reactor compartments in Saida bay, where some part of reactor compartments had already been located, and it will perform supervision in the course of this facility operation in Saida.

According to the RF legislation, all significant projects, including construction of infrastructure for nuclear and radiation-hazardous facilities, are authorized only after environmental impact assessment of this design (OVOS). OVOS with respect to the field of MoD supervision must be approved by the Department.

In the course of OVOS, the comprehensive assessment of arising risk is envisaged. The Department, when considering the design OVOS, evaluates this risk value, as well as puts the appropriate requirements to improve safety and, as necessary, presses for taking additional organizational and technical actions right up to the project forbidden.

Thus, despite the above-mentioned specificity, the Department activity joint with other federal authorities promotes performance of comprehensive regulation and supervision in the field of design, manufacturing, operation and decommissioning of military nuclear powered installations.

Philosophy of the Occupational and Public Radiological Protection in the Regulation of the Nuclear Legacy

Dr. J. Valentin

Abstract This paper describes the applications of the ICRP's recommendations on occupational and public radiological protection in the regulation of the nuclear legacy.

Keywords ICRP, planned exposure situations, emergency exposure situations, emergency exposure situations

1 What Is ICRP?

The International Commission on Radiological Protection, ICRP, was established in 1928 by the International Society of Radiology (ISR). Its mission is to 'advance for the public benefit the science of radiological protection, in particular by providing recommendations and guidance on all aspects of protection against ionising radiation'.

Thus, ICRP is a non-governmental advisory body, offering its recommendations to regulatory and advisory agencies. In addition, ICRP hopes that its advice is of help to management and professional staff with responsibilities for radiological protection. The recommendations and guidance of ICRP are published in the Commission's journal, *Annals of the ICRP*. ICRP has no executive role and does not participate in practical tasks (inspections, remediation, or similar).

The legal seat of ICRP is in the United Kingdom, while its small Scientific Secretariat is currently placed in Sweden. The activities of ICRP are financed mainly by voluntary contributions from national and international bodies with an interest in radiological protection. Some additional funds accrue from royalties on ICRP publications.

Scientific Secretary, ICRP, Sweden

2 ICRP and the Nuclear Legacy

The international Basic Safety Standards of the United Nations [1] are based explicitly on the fundamental Recommendations of ICRP, and national legislation in almost all countries in the world is also based on those Recommendations. Thus, the main contribution of ICRP to the handling of the nuclear legacy is its basic Recommendations.

In addition, some of the more specific guidance of ICRP is particularly relevant for those who are planning or performing actions related to the nuclear legacy. The most pertinent reports include ICRP *Publications 46* [2], *77* [7], and *81* [8] on the disposal of radioactive waste, *Publications 64* ([4] and *76* [6] on protection from potential exposures, *Publication 75* ([5] on protection of workers, and *Publication 82* ([9] on protection of the public from prolonged exposures. Several other reports provide more detailed advice on specific protection issues that may also be relevant.

3 The Fundamental ICRP Recommendations

In March 2007, ICRP approved its revised fundamental Recommendations for a System of Radiological Protection. These will consolidate, develop, and replace formally the previous recommendations issued in 1991 as *Publication 60.* The major features of the ICRP[10].

Recommendations are:

- Updating the radiation and tissue weighting factors in the quantities equivalent and effective dose and updating the radiation detriment based on the latest available scientific information
- Maintaining the Commission's three fundamental principles of radiological protection, namely justification, optimisation, and the application of dose limits, and clarifying how they apply to radiation sources and to individuals
- Abandoning the process-based protection approach using practices and interventions, and moving to a situation-based approach applying the same fundamental principles to all controllable exposure situations, now characterised as planned, emergency, and existing exposure situations
- Maintaining the Commission's individual dose limits for effective dose and equivalent dose from all regulated sources in planned exposure situations
- Re-enforcing the principle of optimisation of protection, which should be applicable in the same way to all exposure situations, with restrictions on individual doses given by dose constraints for planned exposure situations and by reference levels for emergency and existing exposure situations
- Including an approach to a framework to demonstrate radiological protection of non-human species, noting that there is no detailed policy provided at this time

An understanding of the health effects of ionising radiation is central to Commission's recommendations. The distribution of risks to different organs/tissues is judged to have changed somewhat since *Publication 60* ([3], particularly in respect of the risks of breast cancer and heritable disease. However, assuming a linear response at low doses the combined risk of excess cancer and heritable effects remains unchanged at around 5% per Sv, using an unchanged dose and dose-rate effectiveness factor of 2 for solid cancers. The Commission also judges that following prenatal exposure, (a) cancer risk will be similar to that following irradiation in early childhood; and (b) a low dose threshold exists for the induction of malformations and severe mental retardation. The Commission has retained the dose limits for the skin, hands/feet and eye in *Publication 60* but recognises that further scientific reviews and revised judgements may be required in respect of the eye. The available data on excess non-cancer disease (e.g., cardiovascular disorders) are judged to be insufficient to inform on risks at low doses.

The Commission's review of the health effects of radiation has not indicated that any fundamental changes are needed to the system of protection. Existing numerical recommendations in the guidance issued since 1991 remain valid unless otherwise stated.

The central tenet of a linear dose – response relationship for cancer and hereditary effects at low doses continues to provide the basis for the summation of doses from external sources of radiation and intakes of radionuclides. The use of equivalent and effective dose remains unchanged, but the values of radiation weighting factor used for neutrons and protons have been updated. Radiation weighting factors for photons, electrons, muons, and alpha particles are unchanged. Doses from external and internal sources will be calculated using reference 'voxel' phantoms based on medical tomographic images. For adults, equivalent dose coefficients will be calculated by sex-averaging of values obtained using male and female phantoms. Effective dose coefficients will be calculated using sex-averaged equivalent dose coefficients and revised tissue weighting factors, based on updated risk data and intended to apply as rounded values to a population of both sexes and all ages.

Some changes to the structure and terminology of the system of protection were considered desirable in order to improve clarity and utility. In particular the distinction between practices and interventions may not have been clearly understood. Additionally, there were exposure situations which were difficult to categorise in this manner. The Commission now recognises three types of exposure situations which replace the previous categorisation into practices and interventions:

Planned exposure situations involving the planned introduction and operation of sources.

Emergency exposure situations which are unexpected situations that occur during the operation of a planned situation, or from a malicious act, requiring urgent attention.

Existing exposure situations which are exposure situations that already exist when a decision on control has to be taken, including natural background radiation.

The three key principles of radiological protection are retained in the revised recommendations. The principles of justification and optimisation apply in all three exposure situations, whereas the principle of application of dose limits applies only in planned exposure situations. These principles are defined as follows:

The Principle of Justification: Any decision that alters the radiation exposure situation should do more good than harm.

The Principle of Optimisation of Protection: The likelihood of incurring exposure, the number of people exposed and the magnitude of their individual doses should all be kept as low as reasonably achievable, taking into account economic and social factors.

The Principle of the Application of Dose Limits: The total dose to any Individual from regulated sources in planned exposure situations other than medical exposure of patients should not exceed the appropriate limits specified by the Commission.

The Commission continues to distinguish between three categories of exposure: occupational exposures, public exposures and medical exposures of patients.

The revised (2007) Recommendations emphasise the key role of the principle of optimisation. This principle should be applied in the same manner in all exposure situations. Restrictions are applied to individual doses, namely dose constraints for planned exposure situations and reference levels for emergency and existing exposure situations. Options implying doses greater than such restrictions should be rejected. These dose restrictions are applied prospectively. If following the implementation of an optimised protection strategy, the value of the reference level or constraint is exceeded, the reasons should be investigated but this fact alone should not necessarily prompt regulatory action.

Planned exposure situations encompass sources and situations that have been appropriately managed within the Commission's previous recommendations for practices. Protection during medical use of radiation is also included here. Protection in planned exposure situations should take account of deviations from normal operating procedures including accidents and malicious events. Recommendations for planned exposure situations are similar to those provided in *Publication 60* and subsequent ICRP reports.

Emphasis on optimisation using reference levels in emergency and existing exposure situations focuses attention on the expected dose remaining after implementation of protection strategies. This expected dose should be below the selected reference level. These exposure situations often involve multiple exposure pathways which means that protection strategies involving many different protective actions will have to be considered.

Emergency exposure situations include consideration of emergency preparedness and emergency response. Emergency preparedness should include planning for the implementation of optimised protection strategies to reduce exposures to below the selected reference level. During emergency response, the reference level would act as a benchmark for evaluating the protective actions and as an input to the need for further actions.

Existing exposure situations include naturally occurring exposures as well as exposures from past practices conducted outside the Commission's recommendations and past accidents. In this type of situation, protection strategies will often be implemented in an interactive, progressive manner over a number of years.

The revised recommendations acknowledge the importance of protecting the wider environment. To provide a sound framework for environmental protection in all exposure situations, the Commission proposes use of reference animals and plants. In order to establish a basis for acceptability, additional doses calculated to these reference organisms could be compared with doses known to have biological effects and with dose rates normally experienced in the natural environment. The Commission, however, does not propose to set any form of 'dose limits' for environmental protection.

Although the revised recommendations do not contain any fundamental changes to the radiological protection policy, hopefully they will help to clarify application of the system of protection, thereby improving the already high standards of protection.

References

1. IAEA, 1996. International Basic Safety Standards for Protection against Ionizing Radiation and for the Safety of Radiation Sources. Safety Series 115. International Atomic Energy Agency, Vienna.
2. ICRP, 1985. Radiation protection principles for the disposal of solid radioactive waste. ICRP Publication 46. Ann. ICRP 15 (4).
3. ICRP, 1991. 1990 Recommendations of the International Commission on Radiological Protection. ICRP Publication 60. Ann. ICRP 21 (1–3).
4. ICRP, 1993. Protection from potential exposure: a conceptual framework. ICRP Publication 64. Ann. ICRP 23 (1).
5. ICRP, 1997a. General principles for the radiation protection of workers. ICRP Publication 75. Ann. ICRP 27 (1).
6. ICRP, 1997b. Protection from potential exposures: application to selected radiation sources. ICRP Publication 76. Ann. ICRP 27 (2).
7. ICRP, 1997c. Radiological protection policy for the disposal of radioactive waste. ICRP Publication 77. Ann. ICRP 27 (Suppl).
8. ICRP, 1998. Radiation protection recommendations as applied to the disposal of long-lived solid radioactive waste. ICRP Publication 81. Ann. ICRP 28 (4).
9. ICRP, 1999. Protection of the public in situations of prolonged radiation exposure. ICRP Publication 82. Ann. ICRP 29 (1/2).
10. ICRP, 2007. 2007 Recommendations of the International Commission on Radiological Protection. ICRP Publication 103. Ann. ICRP 37 (1/3).

Session III
Challenges in Practical Implementation of Remediation Strategy in Russia and Abroad

Chairmen: Nikolaev, SevRAO
and D.R. Thomas, DTI

Regulation at Hanford – A Case Study

A.R. Hawkins

Abstract Hanford has played a pivotal role in the United States' defense for more than 60 years, beginning with the Manhattan Project in the 1940s. During its history, the Hanford Site has had nine reactors producing plutonium for the United States' nuclear weapons program. All the reactors were located next to the Columbia River and all had associated low-level radioactive and hazardous waste releases. Site cleanup, which formally began in 1989 with the signing of the *Hanford Federal Facility Agreement and Consent Order*, also known as the Tri-Party Agreement, involves more than 1,600 waste sites and burial grounds, and the demolition of more than 1,500 buildings and structures. Currently cleanup is scheduled to be complete by 2035. Regulatory oversight of the cleanup is being performed by the U.S. Environmental Protection Agency (EPA) and the Washington State Department of Ecology (Ecology) under the *Comprehensive Environmental Response, Compensation, and Liability Act* (CERCLA) and the *Revised Code of Washington*, "Hazardous Waste Management."

Cleanup of the waste sites and demolition of the many buildings and structures generates large volumes of contaminated soil, equipment, demolition debris, and other wastes that must be disposed of in a secure manner to prevent further environmental degradation. From a risk perspective, it is essential the cleanup waste be moved to a disposal facility located well away from the Columbia River. The solution was to construct a very large engineered landfill that meets all technical regulatory requirements, on the Hanford Site Central Plateau approximately 10 km from the river and 100 m above groundwater. This landfill, called the Environmental Restoration Disposal Facility or ERDF is a series of cells, each 150 × 300 m wide at the bottom and 20 m deep.

This paper looks at the substantive environmental regulations applied to ERDF, and how the facility is designed to protect the environment and meet regulatory requirements. The paper describes how the U.S. Department of Energy (DOE), EPA, and Ecology interact in its regulation. In addition, the response to a recent $1 million regulatory fine is described to show actual interactions and options in this aspect of the regulatory process.

Hanford National Laboratory, USA

M.K. Sneve, M.F. Kiselev (eds.) *Challenges in Radiation Protection and Nuclear Safety Regulation of the Nuclear Legacy*,

The author acknowledges the significant contributions by Messrs. Clifford Clark and Owen Robertson. Ms. Nancy Williams provided graphics support and Ms. Laurie Kraemer edited the report.

Keywords Hanford site, ERDF, Tri-Party Agreement

1 The Hanford Site

In 1943, the Hanford Site was chosen for plutonium production as part of the Manhattan Project due to its sparse population, remote location, and abundant water supply. The site occupies 586 square miles (1,518 km^2) in Benton County, located in south-central Washington. The Columbia River forms the site's eastern boundary (Fig. 1).

Currently, the Hanford Site is engaged in the largest environmental cleanup effort in the United States. The United States halted plutonium production in the late 1980s when the N-Reactor and PUREX (Plutonium Uranium Extraction) plant ceased operations. In 1989, DOE, EPA, and Ecology signed the *Hanford Federal Facility Agreement and Consent Order* (also called the Tri-Party Agreement). This agreement includes the basic plan and schedule to bring the site into environmental regulatory compliance while cleaning up Hanford's legacy waste (Table 1).

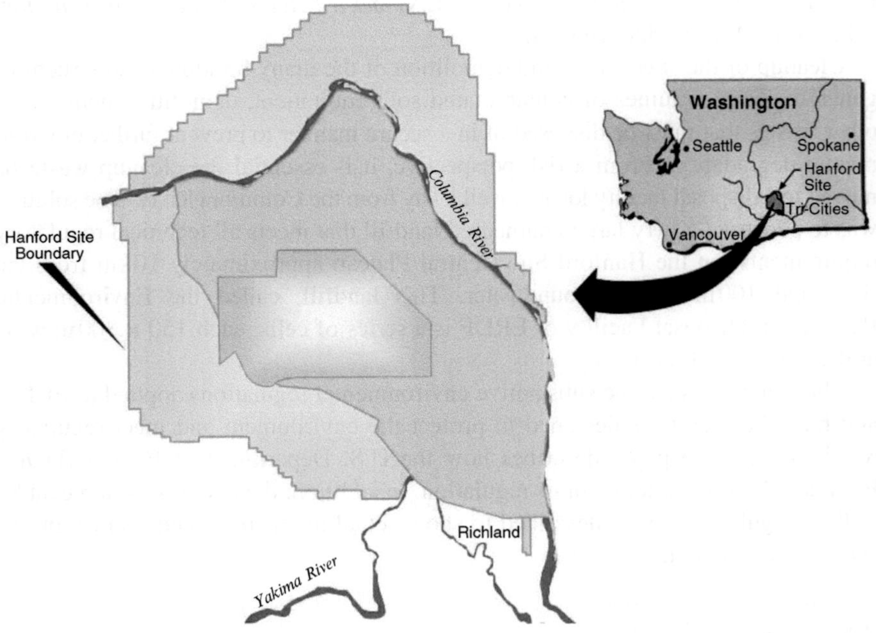

Fig. 1 Hanford site location

Table 1 Legacy material

• Richland Operations Office
• 2,300 t spent nuclear fuel
• Several tons of excess plutonium residues
• About 270 billion gallons of contaminated groundwater, covering about 80 square miles
• About 25 million cubic feet of buried or stored solid waste in 175 waste trenches
• More than 1,600 waste sites and 1,500 facilities (many contaminated), including five processing "canyon" facilities and nine reactor complexes
• Cesium and strontium capsules containing about 109 million curies of radioactivity
• Office of River Protection
• More than 50 million gallons of tank waste in 177 underground storage tanks

DOE has two offices at the Hanford Site overseeing cleanup activities. The Office of River Protection (ORP), established by the U.S. Congress in 1998, manages waste retrieval from, and closure of, 177 underground waste tanks. The ORP also manages the construction of a Waste Treatment and Immobilization Plant that will turn radioactive and chemical wastes into a stable glass form (vitrification). The Richland Operations Office (RL) is responsible for cleaning up the balance of the contamination that is the legacy from the Hanford Site national defense missions. Overall, Hanford cleanup efforts involve more than 11,000 employees and an annual budget of about $2 billion.

RL is focused on two primary goals: (1) restoring the lands in the Columbia River Corridor to a condition where they are suitable for conservation and recreational uses and (2) transitioning the central portion of the Hanford Site – called the Central Plateau because land rises to approximately 300 ft (91 m) above the river – to a modern, protective, waste management operation (Fig. 2).

The River Corridor stretches out over 210 square miles (544 km^2) along a little over 50 miles (80 km) of the Columbia River shoreline. Nine former plutonium production reactors, fuel fabrication sites, research and support facilities, and hundreds of waste sites are located in the River Corridor area. The nine production reactors are being put into interim safety storage condition to be remediated in the future. With few exceptions, the rest of the buildings, structures, and facilities in the River Corridor are being remediated, including all waste sites. The wastes from all the remediation activities are being disposed in the Environmental Restoration Disposal Facility (ERDF).

The Central Plateau covers approximately 75 square miles (194 km^2) formerly dedicated to plutonium recovery operations and managing wastes. The Central Plateau is being transitioned for long-term use to manage, treat, store, and dispose of wastes generated on the plateau and in other areas of the Hanford Site. The Central Plateau contains approximately 1,000 buildings and structures, including five large chemical processing facilities, and 850 waste sites, including the Central Waste Complex, ERDF, and other facilities that are currently being used for waste management and disposal.

In addition to cleanup, a key DOE objective is to shrink the area of the Hanford Site for which DOE is responsible (Fig. 3). The ultimate DOE goal is to release the balance of the land for other uses and possibly for management by another government agency such as the U.S. Fish and Wildlife Service. DOE would only retain responsibility for the waste management activities on the Central Plateau.

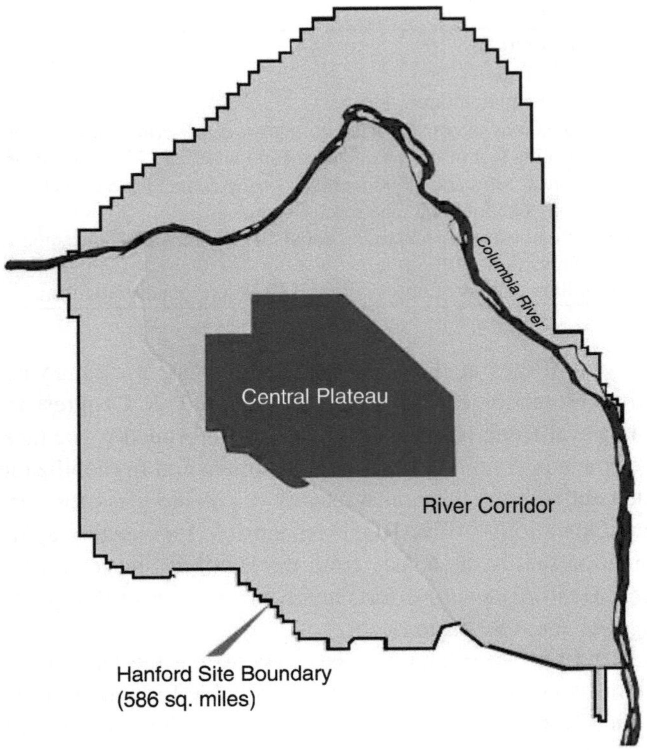

Fig. 2 Areas of cleanup focus

2 Regulatory Environment

The Hanford Site was placed on the *Comprehensive Environmental Response, Compensation, and Liability Act* (CERCLA) National Priorities List (NPL) in 1989. Four sub-areas of the Hanford Site (100 Areas, 200 Areas, 300 Area, and 1,100 Area [Fig. 4]) were officially listed on the NPL on November 3, 1989. The 100 Areas NPL Site and the 300 Area NPL Site are included in the River Corridor Project. The 200 Areas are in the Central Plateau Project. Remediation of the 1,100 Area NPL Site has been completed, and the 1,100 Area has been deleted from the NPL.

In addition, Resource Conservation and Recovery Act of *1976* (RCRA) [7] provisions governing compliance, permitting, closure, and post-closure care of treatment, storage, or disposal (TSD) units apply to active TSD units. The 100 Areas, 200 Areas, and 300 Area are considered active TSD units.

DOE is conducting Hanford Site cleanup in accordance with regulatory requirements under CERCLA, RCRA, the *Atomic Energy Act of 1954* (AEA) [3], Executive Order 12580 (Superfund Implementation) [5], and the *Revised Code of Washington* "Hazardous Waste Management" [2]. The Tri-Party Agreement, signed by DOE, EPA, and Ecology on May 15, 1989, is the legally enforceable agreement

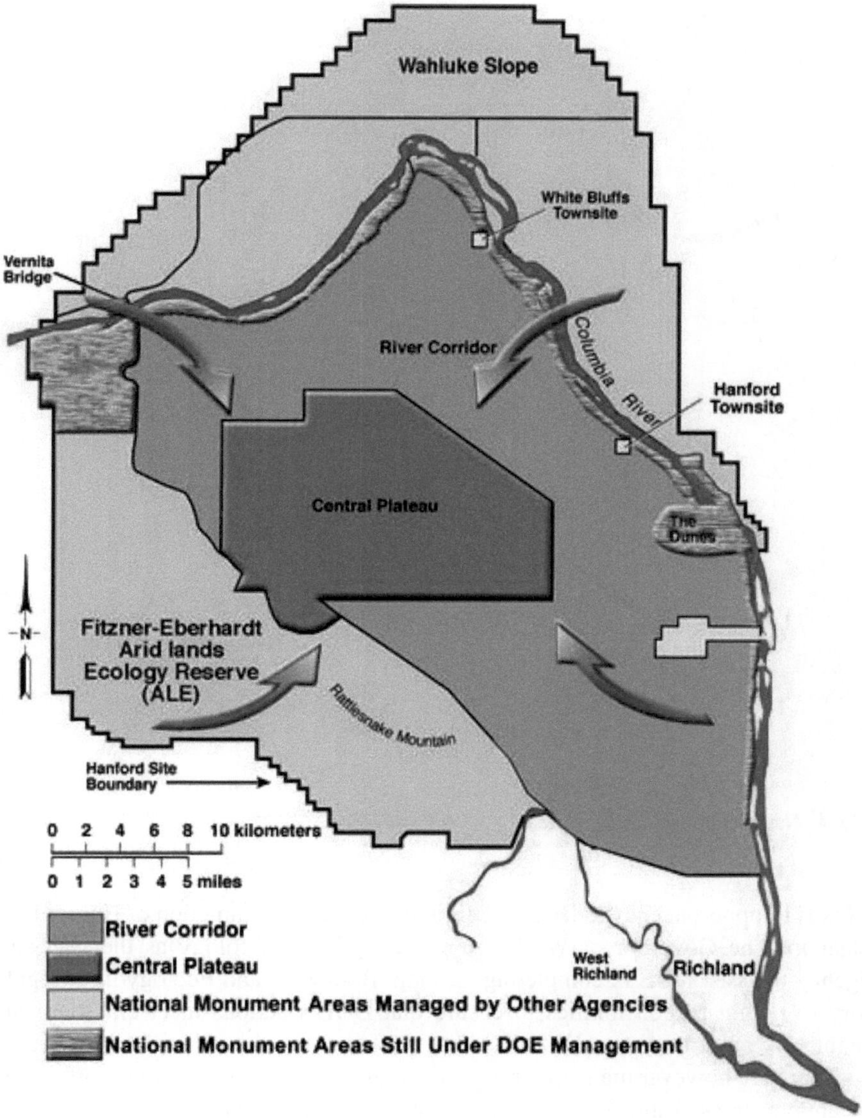

Fig. 3 Shrinking the site

for complying with CERCLA remedial action provisions and with RCRA. The Tri-Party Agreement defines how the three agencies work together to accomplish Hanford Site cleanup, as well as how the agencies interact to meet their individual and collective responsibilities.

DOE is the "Lead Agency" under CERCLA and has ultimate responsibility for completing the remediation of the Hanford Site in compliance with the applicable or relevant and appropriate environmental regulatory requirements. Under CERCLA,

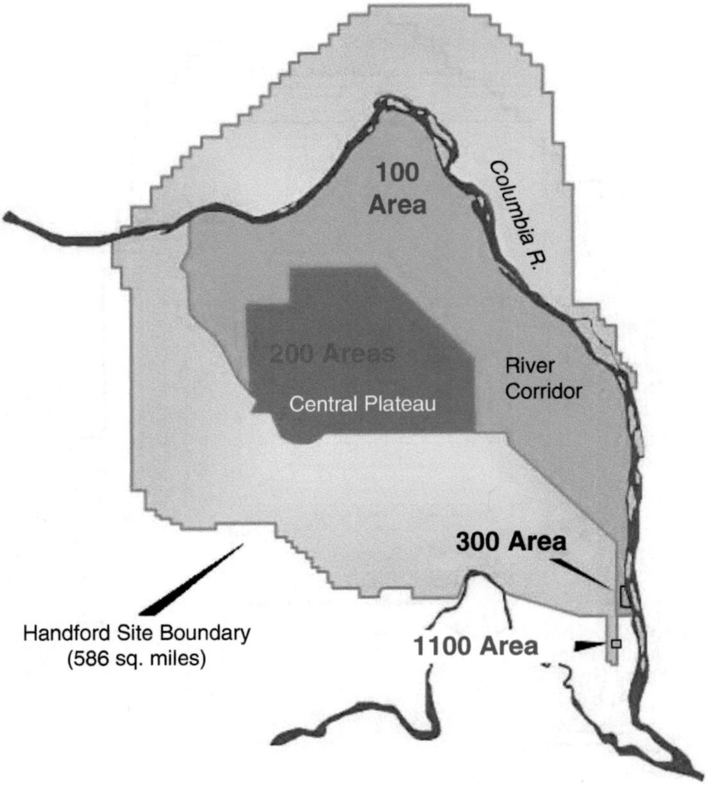

Fig. 4 National priority cleanup areas

EPA is a support agency to DOE to facilitate successful completion of Hanford Site cleanup. The Governor of Washington State selected Ecology as the lead state agency to assist DOE in completing cleanup. Both EPA and Ecology have regulatory oversight responsibilities to ensure that DOE's actions meet environmental regulatory requirements.

There are however many other organizations that have a role in the Hanford Site cleanup process. Table 2 shows these agencies and their roles.

3 Role of the Environmental Restoration Disposal Facility

The cleanup of Hanford's River Corridor would not be possible without ERDF. The landfill, located in the middle of the Hanford Site Central Plateau, was opened in 1996. Without ERDF, wastes would have to be shipped to an offsite disposal facility at a much higher cost. ERDF disposal costs are about $30 per ton, including transportation.

Table 2 Hanford cleanup roles

DOE
– Federal Lead Agency
– Ultimate responsibility for all CERCLA actions
– Responsible to select and execute remedial actions
– Responsible for funding all cleanup actions
– Must reimburse Washington State for all costs to oversee cleanup
– Natural resource trustee
EPA
– Supports DOE in selecting remedial actions
– Must approve selected remedial actions
– Must approve sampling and analysis plans
– Upon request from DOE, delete sites from NPL
– Lead regulatory agency for oversight of CERCLA actions
Department of Interior (US Fish and Wildlife Service)
– Responsible for management of the Hanford Reach National Monument
– Natural resource trustee
– National Oceanic and Atmospheric Administration
– Natural resource trustee (primarily Columbia River)
Washington Department of Ecology
– Regulates RCRA hazardous waste TSD units
– Supports EPA in carrying out its CERCLA responsibilities
– Lead regulatory agency for some remedial actions
– Natural resource trustee
Oregon State
– Natural resource trustee
Nez Perce Tribe, Confederated Tribes of the Umatilla Indian Reservation, Yakama Nation
– Each separately has sovereign nation status
– Each is a Natural resource trustee

Designed to be expanded, ERDF currently consists of six disposal cells. Additional cells are constructed two cells at a time, as needed (Fig. 5).

The surface area on the ERDF floor is 1.5 million square feet (140,000 m^2). As of June 2007, ERDF contained almost 7 million tons of contaminated material from the River Corridor. This includes material from waste sites and burial grounds, as well as demolition debris from hundreds of facilities (Fig. 6).

The ERDF has both a primary and secondary liner system that contains and collects rainwater or water used for dust suppression (Fig. 7). The water that makes its way through the waste and collected by the liner system is called leachate, and contains hazardous and radioactive materials that leach out of the waste, although to date concentrations have been low. The leachate is collected and sent to the onsite Effluent Treatment Facility for evaporation. The resulting solids are returned to ERDF for disposal.

Located under the primary and secondary liner system is a 1-m thick layer of dense clay and native soils that are mixed and compacted to protect the underlying groundwater from a failure in the liner system. The distance between the clay-soil mixture and the groundwater – about 250 ft (75 m) – is an additional, natural barrier.

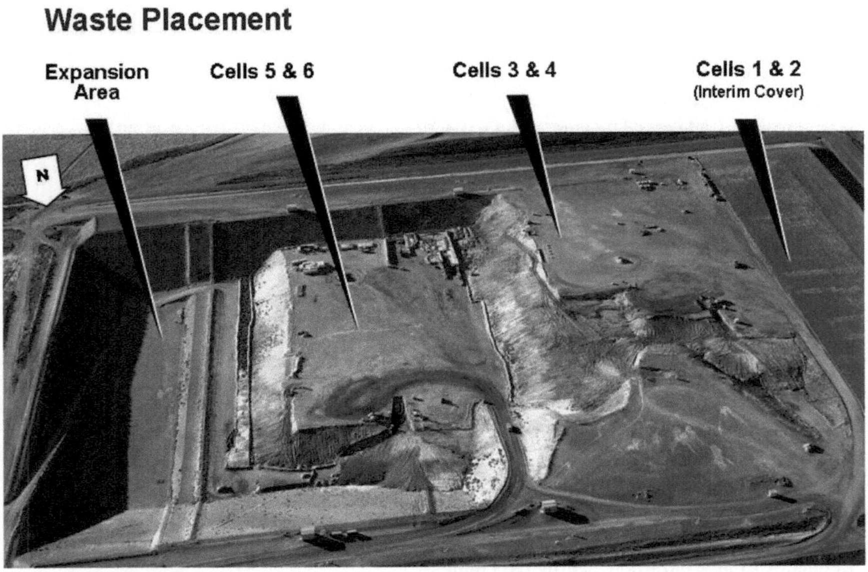

Fig. 5 ERDF design

Fig. 6 Typical debris going to ERDF

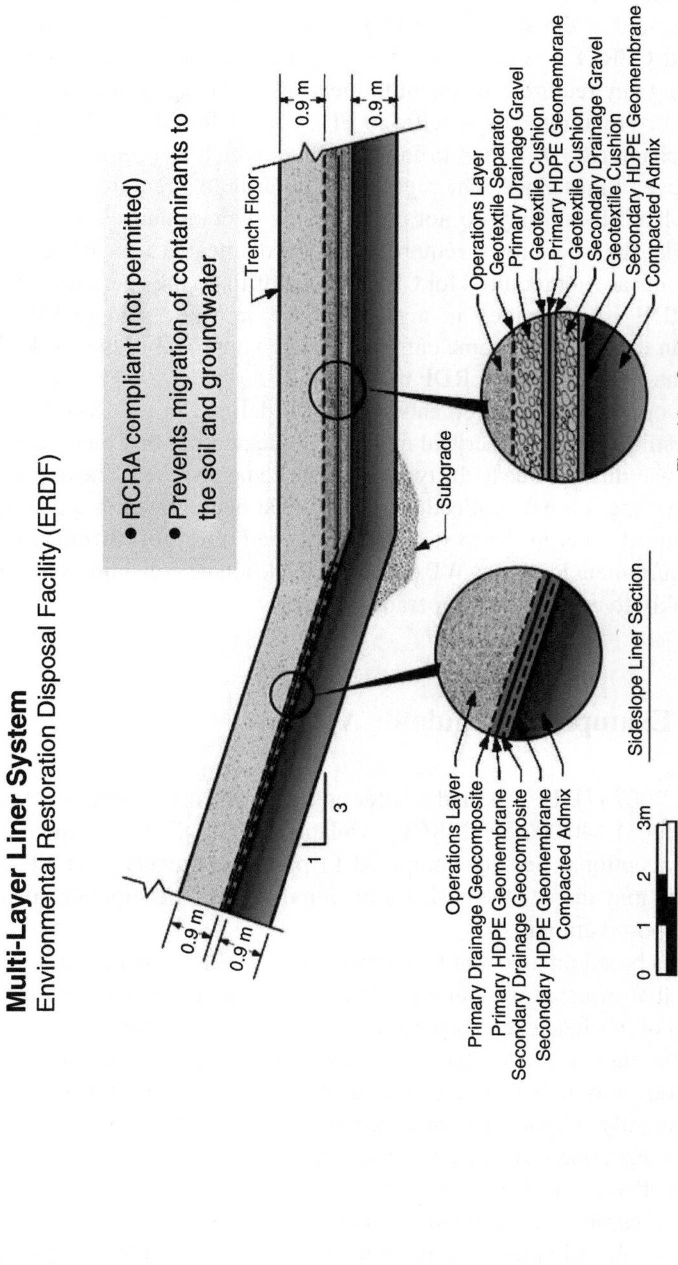

Fig. 7 ERDF protective design liner system

Several years ago, an interim cover was placed on that portion of ERDF that had been filled to capacity. The cover will be expanded as other sections of ERDF are filled. Once the entire facility is full and ready to be closed, a final, permanent cover will be installed.

CERCLA cleanup activities on the Hanford Site began in earnest in 1992. Until ERDF operations began in 1996, wastes from CERCLA removal and remedial actions were stored at the sites where they were generated. Since ERDF became operational, all CERCLA wastes have been disposed in ERDF, either directly or after completing any required treatment to meet ERDF acceptance criteria.

The regulatory requirements specific to ERDF were first defined in the ERDF Record of Decision (ROD) issued in January 1995, which has been supplemented or amended seven times since. The regulations describe the requirements for construction of a landfill but they do not prescribe the operational requirements. For RCRA landfills, the operational requirements are delineated in a RCRA permit. However, permits are not required for CERCLA landfills so the operational requirements for ERDF are described in a remedial action work plan (RAWP). The requirements in the RAWP became enforceable upon approval of the work plan by the lead regulatory agency; for ERDF that is the EPA.

One of the operational requirements for ERDF defined in the RAWP calls for compaction testing using a prescribed method. Implementing that method successfully at ERDF was difficult due to the types of waste being disposed. The requirements also specify that the leachate collection system must be operated in a manner that ensures the liquid level in the system never exceed one foot (30 cm) in depth. A separate requirement in the RAWP called for the leachate collection system to be inspected weekly to ensure it was operating properly.

4 Recent Example of Regulator Action

On March 27, 2007 [1], EPA issued a letter notifying RL it was prepared to assess penalties totaling $1,140,000 for CERCLA violations at ERDF. As shown in Table 3, this was an exceptional penalty compared to past assessments. The EPA letter stated that DOE may dispute the basis for the imposition of the stipulated penalties, but not the proposed amount.

The fine was based on an event (assumed to be a lightning strike) that occurred in May 2006 that affected the pumps that are designed to operate automatically when the level of leachate exceeds prescribed settings. Contractor management did not discover the inoperable leachate pumps in two of the six disposal cells until December 2006, although technicians had recorded the lack of flow from the pumps. Additionally, a management assessment triggered by this event revealed that required compaction tests were not completed for a period of 6 weeks (2 weeks in October and 4 weeks in November 2006).

Additional contractor assessments conducted in response to these findings revealed some compaction test data did not correspond to records of entry into the

Table 3 Recent Hanford site environmental fines and penalties

Amount levied ($)	Date issued	Agency	Penalty basis	Disposition
1,140,000	03/27/07	EPA	CERCLA. Leachate collection system problems; noncompliance with ERDF Operations Plan (compaction)	TBD
120,000	10/16/06	EPA	Noncompliance with CERCLA	Paid
75,000	04/28/05	EPA	Failure to complete Tri-Party Agreement milestone	Paid
270,000	09/21/04	Ecology	RCRA. Shipment noncompliance	Paid
76,000	04/03/03	EPA	Failure to complete Tri-Party Agreement milestone	Paid
57,800	03/26/01	Ecology	RCRA. Improper storage of chemical (collodian)	Closed by settlement agreement

contaminated area where compaction tests are performed. When the technician responsible for taking these tests was confronted with this discrepancy, he admitted to not performing the compaction tests and indicated he had fabricated the data.

The Tri-Party Agreement maximum stipulated penalties are $5,000 for the first week of violation and $10,000 every week thereafter. The EPA calculated its penalty on the basis of: (1) failure to perform weekly inspections that would have detected the presence of leachate and the improper functioning of the leachate system ($305,000 for 31 weeks of violation), and (2) failure to perform compaction tests ($835,000 for 84 weeks of violation). EPA calculated the penalty to the maximum amount based on their belief these were serious and significant violations. (Proper compaction is essential to ensure subsequent waste settlement does not damage the eventual ERDF cap.) While the contractor is corporately responsible for the fine (no government funds are used to pay), RL remains responsible for legal agreements and for satisfying the EPA.

DOE and the contractor have taken extensive corrective action including strengthening disciplined conduct of operations, upgrading work documents, purchasing new compaction equipment, conducting formal lessons learned and increasing oversight. ERDF has largely returned to its former operational status.

5 Mitigating the Fine

Unlike other recent penalties, in this case RL and the contractor, with support from EPA, have proposed implementing supplemental environmental projects (SEP) rather than making a payment to the U.S. Treasury. This option, which can only be used to mitigate a portion of the fine (typically 75%), addresses environmental opportunities and needs in the local area.

DOE and the contractor have proposed two SEPs. The first provides funding to a local university for construction of a greenhouse and nursery facility to be used to produce native plants and seeds that may be beneficial for vegetation of closed/capped waste sites and landfills at the Hanford Site, and for potential revegetation of other areas. The availability of native plant and seed sources for revegetation is currently very limited, both in quantity and number of species. For example, after recent fires, non-native seed was used to restore part of those lands because native seed was not available. Invasive, non-native species are a concern, especially to the Tribal Nations. The expected benefit of this SEP includes creation of a seed bank and improved ability for the local university to teach ecological restoration skills. Figure 8 shows an existing (smaller) university greenhouse currently used to grow native vegetation after which the new greenhouse will be modeled.

The second SEP recognizes two-thirds of Benton County is bordered by the Columbia River. The total perimeter length of the Columbia River around Benton County is approximately 182 miles (292 km). This SEP provides assistance to emergency planning and response organizations to help respond better to oil spills and hazardous substance releases to local rivers by providing two boats to local law enforcement. The SEP includes an agreement to form a response team of personnel from local agencies to respond to emergency incidents, including oil and hazardous substance releases, on the Columbia, Yakima, Walla Walla, and Snake Rivers. Figure 9 shows one of the two boats to be provided.

Fig. 8 Existing university greenhouse

Fig. 9 Emergency response boat

6 Conclusions

Although both the Hanford cleanup and its related regulatory process are complex, cooperation among the TPA agencies and other entities with Hanford interests has supported significant cleanup progress. ERDF continues to play a key role in this progress. Recent issues with disciplined conduct of operations resulted in an unprecedented fine and EPA as the lead regulator withdrawing support for operations. Through broad and proactive response actions ERDF was able to resume waste placement. In addition, DOE, the contractor, and EPA are working to ensure much of the fine is used to support local environmental improvements.

References

1. Letter, to Manager, Richland Operations Office Form Director, Office of Environmental Cleanup EPA, "Stipulated Penalties for Violations of CERCLA Requirements at the Environmental Restoration Disposal Facility," dated, March 27, 2007, U.S. Environmental Protection Agency, Washington, DC.
2. "Hazardous Waste Management," *Revised Code of Washington*, Chapter 70.105, as amended.
3. *Atomic Energy Act of 1954*, Public Law 83–703, 68 Stat. 919, 42 USC 2011.
4. *Comprehensive Environmental Response, Compensation, and Liability Act of 1980*, 42 USC 9601, et seq., as amended.
5. Executive Order 12580, "Superfund Implementation," *Federal Register*, 52 FR 2923, January 29, 1987.
6. *Hanford Federal Facility Agreement and Consent Order*, Washington State Department of Ecology, U.S. Environmental Protection Agency, and U.S. Department of Energy, Olympia, Washington, as amended.
7. *Resource Conservation and Recovery Act of 1976*, 42 USC 6901, et seq., as amended.

Features of Solving the Problems of Remediation of "Sevrao" Facilities: Strategic Planning of Ecological Remediation of the Facility for Spent Nuclear Fuel and Radiation Waste Temporary Storage in Gremikha

B.K. Bylkin[1], Yu.E. Gorlinsky[1], V.A. Kutkov[1], O.A. Nikolsky[1], V.I. Pavlenko[1], Yu.V. Sivintsev[1], B.S. Stepennov[1], and N.K. Shandala[2]

Abstract This article contains principles, criteria and results for selection of the most acceptable end-states of the site and adjacent area after ecological remediation of Radiation Legacy Facility of the former USSR – the Facility for spent nuclear fuel and radiation waste temporary storage in Gremikha (Gremikha TSF). The strategy of end-state reaching has been developed. Based on consideration of existing situation and positions of stakeholders a conceptual scenario of the ecological remediation of Gremikha TSF was proposed. It is based on elaborated strategy and end-state of environmental remediation, and includes four stages:

Stage 1. (2005–2010) – conversion of Gremikha TSF to bring it in correspondence with current normative requirements

Stage 2. (2008–2015) – temporary operation of Gremikha TSF to remove spent nuclear fuel and primary radiation waste from the site

Stage 3. (2010–2020) – decommissioning of Gremikha TSF with removal of secondary radioactive waste generated at it

Stage 4. (2018–....) – remediation of the former Gremikha TSF site

A complex quantitative criteria of the end state of ecological remediation of Gremikha TSF were proposed basing on 2007 ICRP Recommendations. The criteria require in future to be agreed by stakeholders and approved by appropriate supervisory agencies as a possible partial solution.

[1]Russian Research Centre "Kurchatov Institute", 1, Kurchatov square, Moscow, 123182, Russian Federation

[2]State Scientific Centre "Institute of Biophysics", 46, Zhivopisnaya street, Moscow, 123182, Russian Federation

For correspondence or reprint contact: V. Kutkov, Russian Research Centre "Kurchatov Institute" (RRC KI), 1, Kurchatov square, Moscow, 123182, Russian Federation. Tel: + 7 (499) 1967354, + 7 (499) 1969152; Fax: + 7 (499) 1968673, or email Kutkov@front.ru.

M.K. Sneve, M.F. Kiselev (eds.) *Challenges in Radiation Protection and Nuclear Safety Regulation of the Nuclear Legacy*,

Keywords End-states, adjacent area, ecological remediation, Gremikha, radioactive waste management, *Hazard*, *Harm*, ICRP, solid radioactive waste (SRW), **Immediate dismantling (DECON)**, **Delayed dismantling (SAFTOR)**, **On-site disposal (ENTOMB)**

1 Introduction

Objective of this work – based on analysis of existing situation, experience in remediation of contaminated areas of radiation legacy facilities, to develop principles and criteria and propose the most acceptable strategy and scenario for remediation of the site of the Facility for spent nuclear fuel and radiation waste temporary storage in Gremikha (Gremikha TSF). This facility is the Branch of SevRAO, the Federal state unitary enterprise for radioactive waste management in the Northwest region. The SevRAO was organized in 2002 in order to manage former USSR Navy bases at Kola Peninsula operated for many years as storages of spent nuclear fuel and radiation waste. Results of the work are necessary for selection of the site and adjacent area end-state as a result of the ecological remediation of Gremikha TSF and strategy of its reaching.

The objective of the ecological remediation of the Gremikha TSF is spent nuclear fuel (SNF) and radiation waste (RW) removal from its site and subsequent decommissioning this facility together with remediation of the adjacent area within the Sanitary buffer zone accompanied with its cleanup up to acceptable levels of the residual contamination. This activity should be carried out having in mind possible options of further usage of the facility, ensuring, at the same time, safety of workers, the public and the environment.

Some circumstances make difficulties for the environmental remediation of Gremikha TSF. The situation analysis revealed more relevant of them [3, 4]. These circumstances are the following:

- A lack of requirements for Nuclear and Radiation Hazardous Facility (NRHF) ecological remediation and criteria specifying its reaching, in the national normative basis.
- A lack of approved solutions relating to the Gremikha TSF end-state and time of its reaching considering abilities and proposals regarding further usage of its separates sites.
- Gremikha TSF is “radiation legacy” of the former USSR. It was built and used according to the special Norms and Regulations of radiation safety assurance at Defense Ministry’s and Navy’s facilities in the USSR. Their requirements essentially differed from requirements of regulatory control for civil facilities.
- Necessity of hard urgent work implementation in order to set regulatory control of NRHF, under conditions of deterioration of the protective barriers and infrastructure, as well as decreasing of the people-ware of these operations.
- Dependence of the plans of ecological remediation of the Gremikha TSF upon tendencies and perspectives of further development of closed administrative-territorial

formation (CATF) Ostrovnoj – the city, which is the main basis of infrastructure and people-ware for this work.

- Diversity of positions and different objectives of parties interested in further perspective of CATF Ostrovnoj and Gremikha TSF site and ways of reaching end objectives.
- Necessity to considerate interests of donors to ecological remediation works at the Gremikha TSF.
- Seasonal restriction of active work due to natural climatic Northern conditions and absence of land-based communication with the "mainland".
- Specificity of facilities and composition of works on their environmental remediation; complexity of work planning at lack of information about their conditions, obvious multi-choice of administrative and technological solutions and lack of the ecological remediation experience for such facilities.
- Necessity of decision making considering real dynamics of work and instability of conditions of its performance; high risk of successful implementing the planned decisions being made are also to be taken into account.

Development of conceptual technological solutions regarding the environmental remediation of the Gremikha TSF considering above mentioned circumstances is based on the preliminary development of principles and criteria of its accomplishment together with selection of the most acceptable end-state options and strategy of its reaching.

2 Objective Task Description for Ecological Remediation of the Radiation Legacy Facility

2.1 Objective of Ecological Remediation of the Radiation Legacy Facility

Ecological remediation (ER) of the Radiation Legacy Facility (RLF) is a complex of administrative and technical measures on conversion and decommissioning RLF and remediation of the site for RLF location according to its end-state. The objective task of ecological remediation of the Radiation Legacy Facility is elimination of further influence of factors defined RLF as a nuclear and radiation hazardous facility and cleanup of the site and adjacent area up to the levels of the residual contamination acceptable for stakeholders.

The main difference between the ecological remediation of RLF and similar activity for decommissioning and remediation of the active NRHF site is:

- Design, making and completed operation of RLF were performed without consideration of need in ecological remediation of the facility site and adjacent area when its life cycle is finished.
- Design, making and completed operation of RLF were performed in accordance with regulatory requirements that don't correspond to actual requirements to NRHF allocation and operation.

So the most important stage of the ecological remediation of RLF is its conversion into NRHF that corresponds to requirements of actual nuclear and radiation safety regulation. This conversion precedes to removal of nuclear and radioactive materials from the RLF site.

A base for selection of the RLF end-state after completion of its environmental remediation, strategy and criteria for its reaching is consideration of all stakeholder positions. Taking into account these stakeholder opinions is also a base for optimization of radiation protection of workers, the public and the environment at ecological remediation [11].

2.2 Principles of Ecological Remediation of the Radiation Legacy Facility

In terms of IAEA recommendation, the following main **principles of work implementation** with respect to the ecological remediation of RLF are adopted:

- Subdivision of the whole work package into **stages**, characterizing by continuity of their implementation and end-state, which is in compliance with the required safety level for workers, the public and the environment; such subdivision can be established for the longer time period
- Successive reducing level the facility hazard relating to workers, the public and the environment in the course of work implementation
- Consideration of stakeholder positions not only in the course of decision making regarding the facility end-state and its further use, but also during selection of a strategy for its reaching
- Determination of urgent operations directed to putting the conditions of the nuclear and radioactive materials management into compliance with active requirements to NRHF
- Top-priority (as the preliminary work is carried out) removal of nuclear and radioactive materials from the facility site
- Successive reducing contamination level and area and minimization of secondary RW generation in the course of the environmental remediation of RLF, provision of their interim storage in passive safe condition
- Optimization of radiation protection of workers, the public and the environment
- Systematic analysis of the results and correction of the strategy being selected in terms of varying external circumstances and conditions of their performance

So, ecological remediation of the Radiation Legacy Facility includes:

Stage 1 Conversion of RLF – putting its condition into compliance with active requirements of nuclear and radiation safety regulation

Stage 2 Temporary operation of RLF for removal of accumulated nuclear and radioactive materials form the facility site

Stage 3 Decommissioning the RLF – implementation of administrative and technical measures eliminating further use of the facility for its intended purposes and maintenance protection of workers, the public and the environment

Stage 4 Remediation of the RLF **site** and adjacent area within the Sanitary buffer zone – cleanup of its site up to the levels of the residual contamination corresponding to the facility end-state after its decommissioning

Levels of acceptable residual contamination should be set to maintain protection and safety of workers, the public and the environment depending on the option of further use of the site. Levels of residual contamination should be optimized.

2.3 End-State of Radiation Legacy Facility After Its Environmental Remediation

Objective of the ecological remediation of RLF is certain facility end-state when reaching of which remediation works should be finished. As possible RLF end-states after completion of ecological remediation "Brown field" and "Green field" are considered.

Brown field is a state of the site and adjacent area when it's possible to restrictedly use it as:

1. A site for new radiation hazardous facility or
2. A site for facility of other general industrial use

Green field is a state of the site and adjacent area when it's possible to unrestrictedly use it.

These states are characterized by different level of residual contamination of environments which successively reduces while approaching to the state of clear site, the Green field.

The way of reaching certain end-state of remediation subject is determined by the ecological remediation strategy. Reaching specified above RLF end-states after completion of ecological remediation is possible at the following strategies.

1. Function change of RLF – an option of the ecological remediation providing:
   - Facility decommissioning
   - Partial stage-by-stage dismantling (immediate or suspended) and partial liquidation of equipment, systems, structures and structural facilities
   - Removal of all radioactive waste from the RLF site and
   - Remediation of the RLF site for further use as new radiation hazardous facility
2. On site disposal of RLF – an option of the ecological remediation providing:
   - Facility decommissioning
   - Full stage-by-stage dismantling (immediate or suspended) and liquidation of equipment, systems, structures and structural facilities
   - Disposal of secondary radioactive waste at the RLF site
   - Operation of facility as a RW disposal facility – new radiation hazardous facility

3. Termination of RLF – an option of the ecological remediation providing:

   - Facility decommissioning.
   - Full stage-by-stage dismantling (immediate or suspended) and liquidation of equipment, systems, structures and structural facilities.
   - Removal of all radioactive waste from the RLF site, including secondary ones.
   - Remediation of the RLF site for further use as new non-radiation hazardous facility or clear site depending on the quality of decontamination area cleanup.
   - License termination.
   - It's also necessary to have a strategy to maintain radiation protection and safety of workers and the public at the environmental remediation of RLF. Following it would allow:

     – To define radiation protection criteria (dose constrains) for work planning
     – To define radiation protection criteria (dose reference levels) for work performance
     – To define criteria for evaluation of radiation safety during performance of planned works and at their completion

   - Regulatory documents [14, 31, 32], in force at present time in the Russian Federation consider two categories of radiation sources as the radiation safety regulation subjects:

     – Man-caused radiation sources specially built for their useful use or being by-products of such activity
     – Natural radiation sources to which Norms and Regulations of radiation safety assurance are applied

Using a term "Source of radiation" or simply "a source" we follow usual practice of radiation protection and safety. According to it [6], "a source" is radioactive substance or device that emits or can emit ionizing radiation to which Norms and Regulations are applied. As a source all objects that can cause exposure at emission of ionizing radiation or at effluent of radioactive substances and materials are considered. For purposes of radiation protection and safety any radiation source is considered as a source of harm and hazard. Source harm is caused by related actual exposure. Source hazard – by potential exposure, that when the source becomes beyond control could be realized as emergency exposure which causes significant radiation effects [14, 16]. This harm-hazard dualism of a source is the basic characteristic for development source-specific criteria limiting exposure.

Hazard is the basic feature of man-caused radiation sources. Usually such source is specially designed in the form of device or installation where the most effective use of generated ionizing radiation is reached. So such source is initially radiation hazardous and their safety is based on defense in depth concept. Regulatory requirements for management of man-caused source are aimed, first of all, to providing control of such specially built source to prevent its runaway [14, 15, 16]. At that, results of dose control for the public and workers exposure induced by considered source show its safe condition and reliability of defense in depth system [15, 16].

Therefore, the dose limit for public exposure related to man-caused source is very restrictive (1 mSv per annum) to ensure safe condition of the source.

Harm is the basic feature of natural radiation sources. Influence of such source on a person is prolonged and extended in time. Because of low specific activity of natural radioactive materials, their runaway can't cause significant emergency exposure doses. Therefore, the permissible dose related to natural source of public exposure is high and is about 10 mSv per annum [6, 31, 37].

Beside the sources described above there are radiation sources of the third type. These are radioactive contaminations distributed in the environment because of stopped practical activity. As a rule, there are no persons responsible for such contamination and sources themselves are associated with "radiation legacy".

Exposure of the involved public and workers related to "radiation legacy" among others includes:

- Exposure induced by global precipitations of radionuclides because of nuclear weapon tests
- Exposure at contaminated areas of stopped facilities whose condition doesn't correspond to requirements specified for civil radiation hazardous facilities

Exposure induced by such sources is not regulated by national Norms and Regulations while sources themselves are burden that should be eliminated. At present key issues of radiation safety assurance at liquidation of "radiation legacy" of the former USSR are still not solved. This circumstance essentially affects on estimation of acceptability for some end-states of the ecological remediation subjects. Presence of exposure components that don't fall under national normative basis and are not regulated is a problem for radiation safety assurance at liquidation of "radiation legacy" and need to be solved certainly [1, 34, 35]. A lack of this problem solution based on the principles of human radiation protection leads to solutions based on the principles of social protection of the public living at contaminated areas [19, 20, 21]. This fact causes many negative consequences.

Restriction of "radiation legacy" influence on the public is not a new issue [17, 18]. ICRP dedicated to this issue a special publication [2, 10]. Recommendations contained there became a base for Safety Guide issued by IAEA in 2006 [9].

ICRP considers three exposure situations and types:

- Situations of planned exposure, which include exposure situations under controlled conditions of radiation sources management.
- Situations of emergency exposure, which include exposure situations under conditions of uncontrolled exposure induced by sources, which were under regulatory control, but became beyond control following radiation accident.
- Situations of existing exposure, which include situations of exposure with unregulated sources distributed in the environment. These sources have already existed by the moment of decision making regarding necessity of their control. Such situations include exposure due to natural radionuclides, being distributed in the environment, and contamination following radiation accidents and past activity.

Radiation situations in the course of the conversion strategies, change or disposition of function of Gremikha TSF can be referred to three types:

- Planned exposure due to work implementation, dealing with restoration of regulatory control over SNF and RW
- Planned exposure due to the facility operation after change of its functions
- Existing exposure due to residual radioactive contamination of location of new radiation-hazardous facility or industrial facility

In cases of planned exposure ICRP uses several approaches to radiation safety assurance with concept of dose limit, dose constraint and reference level.

Three-component structure of radiation protection regulation levels as a whole corresponds to Russian system. Here quota pretends to the role of dose constraint. Quota is "a part of dose limit specified for limitation of the public exposure induced by particular man-caused radiation source and way of exposure (external, intake with water, food and air)" [31]. Quota and dose constraint are similar but not equivalent on their essence. The quota is always a part of dose limit, but in line with the ICRP, a value of dose could be defined in range 0.01–100 mSv per annum (effective dose) depending on exposure situation and source characteristics [12, 13]. In situation of existing exposure related to natural radioactive materials residual contamination of human environment as a result of completed practice (radiation legacy) ICRP recommends to select reference levels for intervention in case of long-term exposure within the range of 1–20 mSv per year (effective dose) depending on the situation.

3 General Characteristic of the Facility for Spent Nuclear Fuel and Radiation Waste Temporary Storage in Gremikha

Gremikha TSF is a radiation legacy object of the former USSR subject to ecological remediation at present time. It includes 32 process buildings and structures. Layout of basic buildings and structures at the site of Gremikha TSF is presented in Fig. 1. Here buildings 1, 1A and 1B are SNF storages, Building for refueling reactors with liquid metal coolant and Spent core storage respectively; buildings 17 and 19 – Liquid radiation waste (LRW) treatment building and Building for storage of LRW concentrate.

A source of nuclear and radiation hazard at the site of Gremikha TSF is spent nuclear fuel from nuclear submarines with water-water and reactors with liquid metal coolant and RW.

Territory and buildings of the Gremikha TSF is contaminated with ^{60}Co, ^{90}Sr + ^{90}Y, ^{137}Cs + ^{137m}Ba. Local hot spots in buildings and at site are also contaminated with ^{152}Eu, ^{154}Eu; and ^{238}Pu, ^{239}Pu and ^{241}Am. Distribution of ^{137}Cs on soil surface is shown in Fig. 1 together with borders of Sanitary buffer zone (SBZ), Free access zone (FAZ) and Controlled access zone (CAZ) around some facilities.

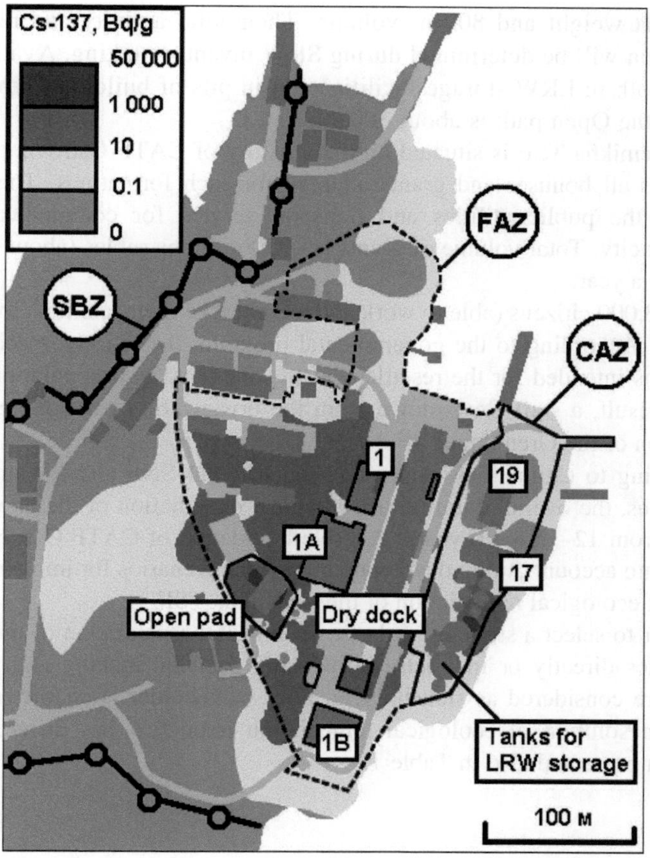

Fig. 1 Layout of buildings and structures at the Gremikha TSF with locations of Controlled and Free access zones in connection with site contamination: solid line – border of Controlled access zones around separate facilities; dashed line – border of Free access zone (safety perimeter of the site) of the Gremikha TSF; solid line with circles – border of Sanitary buffer zone of the Gremikha TSF

Security perimeter of the Gremikha TSF consists wit the outer border of FAZ and the inner border of SBZ. In order to Sanitary regulations of the Russian Federation [32] any practice of land use inside SBZ of an enterprise is prohibited if it does not relate to the operation of the enterprise. The territory of the SBZ could not be inhabited. Annual dose of exposure of members of the public at outer border of SBZ should not exceed 1 mSv per year on average over any successive 5 years, but no more than 5 mSv per one single year; the exposure of workers at workplaces inside SBZ should not exceed 5 mSv per year on average over any successive 5 years, but no more than 12 mSv per one single year.

The main primary solid radioactive waste (SRW) stream is located within the Open Pad, in building 19 and on the site between building 1B and a Dry dock.

It has 450 t weight and 800 m^3 volume. Their total activity and radionuclide composition will be determined during SRW inventory taking. Available LRW volume, both in LRW storage facilities and in pits of building 1, in SNF containers at the Open pad, is about 300 m^3.

The Gremikha TSF is situated on the territory of CATF Ostrovnoj. Now, this CATF gets all bonuses and grants adopted for such formations. Their majority relates to the public utilities and transport service for communications with Murmansk city. Total volume of grants is ~350 millions rubles (about 10 millions Euros) per a year.

About 3,000 citizens (able to work and members of their families) live in CATF Ostrovnoj. According to the governmental program, the citizens receive periodically means intended for the resettlement to "mainland", so population decreases and as a result, a staff flows-down from the organization performing ecological remediation of the Gremikha TSF.

According to estimations, depending on different SNF and RW management technologies, the overall duration of ecological remediation of the Gremikha TSF could be from 12–15 to 20 years. Possible liquidation of CATF Ostrovnoj should be taken into account during development of work scenarios for implementation of strategy of ecological remediation of the Gremikha TSF.

In order to select a strategy of the environmental remediation of the Gremikha TSF, parties directly or indirectly influencing decision making and progress of works were considered as stakeholders. Each stakeholder is characterized by its position in solution of ecological remediation issue that has different form of expression presented also in Table 1.

Table 1 Stakeholders interested in the ecological remediation of Gremikha TSF

Stakeholder	Form of position expression
Representative authorities of the Russian Federation	Laws, directives, policies/concepts; hearing materials etc.
The Russian Federation Government (Rosatom, Ministry of Defense and others)	Leader orders, resolutions of collective authorities, concepts etc.
Regional and local authorities; CATF Ostrovnoj administration and citizens	Resolutions of the local authorities and expert judgments. Materials of the public polls, publications in the central and local press, Internet
Federal and local supervision authorities of Russia	Normative documents, recommendations based on check results and expert judgments
Russian contractors on ER	Expert judgments
Donors and foreign subcontractors on ER	Materials of Contact Expert Group. National and corporative policies, frame or indicative programs, international agreements, contracts etc.

4 Selection of the Most Acceptable End-State of Gremikha TSF and Strategies of Its Attainment

This section presents overall approach to development of a strategy of environmental remediation of Gremikha TSF. It involves:

- Definition of end-states of object of remediation
- Impartial assessment of an acceptability of such end-state and
- Strategy of their attainment

Result of overall approach includes:

1. List of most acceptable end-states of Gremikha TSF after completing environmental remediation and
2. Strategies of their attainment

It is grounded on an all-round evaluation of an acceptability of end-states of remediation for all stakeholders. Before carrying out of this all-round evaluation, definition of such end-state can grows out only administrative decision. Such statement is a starting point for the beginning the search for development of ecological remediation strategy based on the analysis of an acceptability of each option using methodology of multiple factor analysis of expert appraisals.

4.1 Involving Stakeholders in Estimation of End-State Acceptability

Selection of the most acceptable strategy was carried out on the base of analysis of expert appraisal of factors for the end-state options concerned and consideration of pathways of the end-state reaching. Independent experts made these appraisals having in mind the stakeholder positions presented in the relevant documents [1, 3, 4].

4.2 Factors of Acceptable End-State Selection

When selecting the most acceptable option of the strategy of ecological remediation of Gremikha TSF, the following factors are to be taken into account:

1. Compliance with the actual legislation, national policy and international obligations
2. Compliance with ER implementation principles
3. Effectiveness of the protective barriers in the light of their deterioration
4. Hazard and harm of operations for workers
5. Hazard and harm of work for the public and the environment
6. Hazard and harm of the facility end-state for workers
7. Hazard and harm of the facility end-state for the public

8. Ability of ER subdivision into continuous stages with progressive reducing hazard of the facility interim conditions
9. Means providing radiation and nuclear safety
10. Means providing the facility physical protection
11. Availability of required infrastructure, technical support and experience
12. Volume and activity of the secondary waste being generated in the course of work
13. Complexity of SNF and RW management and their removal from the facility site
14. An ability of the facility usage in its end-state
15. Ability of re-using materials (waste)
16. Availability of high qualified staff
17. Labor expenditures, financial costs and work duration
18. Social consequences (compensation and payments) and social provision

General strategy for reaching end goals of ecological remediation can be presented in form of step-by-step algorithm.

Step 1. Selection of the most acceptable end-states of the Gremikha TSF as the subject to ecological remediation and development of strategies of their attainment.

Step 2. Conversion of the Gremikha TSF to bring it into correspondence with current regulatory requirements.

Step 3. Temporary operation of the Gremikha TSF to provide for SNF and RW removal from the site.

Step 4. Making a final decision regarding the end-state of the Gremikha TSF as the subject to ecological remediation and development of strategies of their attainment and normative and legislative support of such decision implementation.

Step 5. Decommissioning of the Gremikha TSF according to a certain end-state of the Gremikha TSF as the subject to ecological remediation.

Step 6. Remediation of the site and adjacent areas according to a certain end-state of the Gremikha TSF as the subject to ecological remediation.

Initial variance of the end-state of Gremikha TSF site as the subject to ecological remediation requires considering as equal (as a result of possible directive decision) three end-states until final decision making:

- Brown field: A site for the radiation hazardous facility
- Brown field: A site for the industrial (non-radiation hazardous) facility
- SNF and RW storage and/or disposal facility for the secondary RW

End-state options determine a strategy of the environmental remediation of Gremikha TSF. They are presented in Table 2 and differ by the level of regulatory control over the ecological remediation subject at its end-state. The strictest regulatory requirements are specified for SNF and RW storage facilities and facilities for RW disposal. The facility in the state of "Green field" characterizes its full release from the regulatory control. Table 2 contains also possible strategies leading to considered end-states.

Table 2 Strategies of the environmental remediation of Gremikha TSF

End-state	Requirements for safety level	Strategy and its meaning
E0: Present state E1: SNF and RW storage and/or disposal facility for the secondary RW	Full compliance with requirements for NRHF and for allocation and operation of SNF and RW storage or disposal facilities	S1: Non-taking actions. Maintenance of the facility in its current state without any changes S2: Conversion and temporary operation of the facility. The facility conversion and its use for its actual purpose for SNF and RW management and for their interim (process) store S3: On site disposal of the facility. The facility conversion, SNF and RW removal, decommissioning and disposal of the rest materials with fixed contamination (secondary RW) *in-situ* (ENTOMB)
E2.1: Brown field: A site for the radiation hazardous facility	Full compliance with requirements for the radiation hazardous facility	S4.1: Change of the facility function. SNF and RW removal and subsequent immediate partial dismantling the buildings, together with decontamination of the building and the area within the SBZ in order to build new radiation hazardous facility (DECON) S4.2: Change of the facility function. The same as S4.1, but with time-delayed partial dismantling (SAFSTOR)
E2.2: Brown field: A site for the industrial (non-radiation hazardous) facility	Release from the regulatory control due to restrictions of the site usage	S5.1: Liquidation of the facility and remediation of its site. SNF and RW removal and subsequent immediate full dismantling the building and RW removal together with remediation of the adjacent area within the SBZ in order to build new non-radiation hazardous facility (DECON) S5.2: Liquidation of the facility and remediation of its site. The same as S5.1, but with time-delayed full dismantling (SAFSTOR)
E3: Greenfield	Full release from the regulatory control	S6.1: The facility liquidation and the site renovation. SNF and RW removal and subsequent immediate full dismantling the building and RW removal together with remediation of the adjacent area within the SBZ (DECON)

(continued)

Table 2 (continued)

End-state	Requirements for safety level	Strategy and its meaning
		S6.2: The facility liquidation and the site renovation. The same as S6.1, but with time-delayed full dismantling (SAFSTOR)

Three main approaches to dismantling the facilities and installations are considered:

1. **Immediate dismantling (DECON)** – Immediate dismantle of the devices, removal of active components and structures from a site of installation.
2. **Delayed dismantling (SAFTOR)** – The spread in time reduction of activity of high radioactive installation elements after the termination of its operation up to a level, allowing to make its safe dismantle (delay period depends on the source term and reach 100 years for installations contaminated by the fission products, e.g. Cs-137).
3. **On-site disposal (ENTOMB)** – The conclusion of radioactive structures in concrete or other long-living material, maintenance service and monitoring of these structures until radiation will not decrease to a level admitting the manipulation with them without what or restrictions.

An acceptable strategy of the ecological remediation of Gremikha TSF was selected for its the most potentially hazardous facilities, such as:

1. The SRW temporary storage site (Open pad), which is in the state of final shutdown
2. Building 1 which is in the state of final shutdown
3. Operating site of LRW storage

Operating site of LRW storage has been considered as a subject of exposure to the water area and possible use for interim storage of secondary LRW in the course of ecological remediation work implementation.

According to results of multifactor analysis, for the environmental remediation of Gremikha TSF, the following end-states and strategies of their attainment from Table 2 were considered as the most acceptable:

- Rank I The most appropriate end-state – E2.1 with strategy S4.1
- Rank II–III Appropriate end-state – E2.2 with strategy S5.1
- Rank II–III Appropriate end-state – E1 with strategy S3

The end-state **E2.1** can be reached in the frames of the active normative and legislative basis. Implementation of the strategies **S5.1** and **S3** requires special decisions regarding their normative and legislative support.

4.3 Numeric Values of the Key Ecological Remediation Criteria

The radiation situation during conversion of the Gremikha TSF and implementation of strategies for function change (**S4.1**) or liquidation (**S5.1**) of its facilities could be referred to two types:

1. Situations of planned exposure due to work implementation, dealing with restoration of regulatory control over nuclear and radioactive hazardous materials
2. Situations of planned exposure due to facility operation after change of its function at implementation of the strategy **S4.1**
3. Situations of existing exposure due to residual radioactive contamination of location of new radiation hazardous facility at implementation of the strategy **S4.1**; or industrial facility at implementation of the strategy **S5.1**

According to new ICRP Recommendations [12, 13], in order to plan some actions directed to radiation protection of workers at the stage of conversion, temporary operation and decommissioning of facilities, it is expedient to set the following dose constraint levels:

1. For planning routine operations, according to new ICRP Recommendations [12, 13], dose constraint could be set at the level of no more than 5 mSv per year.
2. For planning extraordinary operations, according to new ICRP Recommendations [1, 12, 13], dose constraint could be set at the level of no more than 25 mSv per operation.

To assess radiation safety conditions at these stages of the environmental remediation of Gremikha TSF, dose limits from NRB-99 [31] and dose constrains (quotas) from SPORO-2002 [33] should be used for exposure induced by man-caused sources:

1. Occupational exposure induced by all types of sources in the course of routine operations on the site – 20 mSv per year on average over any successive 5 years, but no more than 50 mSv per one single year
2. Public exposure due to handling RW existing at Gremikha TSF – 0.1 mSv per year on average over any successive 5 years, but no more than 0.5 mSv per one single year
3. The Public exposure due to all sources connected with the site – 1 mSv per year on average over any successive 5 years, but no more than 5 mSv per one single year

According to new ICRP Recommendations [12, 13], the objectives of the facility remediation can be considered as achieved, if exposure doses due to the residual contamination do not exceed controlled levels adopted by the National regulatory authority. Their values should be set depending on the facility end-state after the environmental remediation of Gremikha TSF site [1]:

1. If the strategy "On site disposal of Gremikha TSF" (**S3**) is implemented, criteria from NRB-99 are necessary and sufficient for planning measures for radiation protection of the public and workers.

2. If the strategy "Change of Gremikha TSF function" (**S4.1**) is implemented and a site and structures are continued to use for building radiation hazardous facility, the following values for dose reference levels from residual contamination are proposed:

   (a) On workplaces of the personnel from A group – 3 mSv per year
   (b) On workplaces of the personnel from B group – 1 mSv per year
   (c) For the public living within the supervised area – 0.1 mSv per year

Design of such facility and its operation should provide conditions, when annual exposure doses due to operation of this facility do not exceed the following dose constraint levels:

- For the personnel from A group – 7 mSv per year
- For the personnel from B group – 1 mSv per year
- For the public living within the supervised area – 0.15 mSv per year due to release of radioactive materials from the site

If the strategy "Liquidation of Gremikha TSF and remediation of its site" (**S5.1**) is implemented and a site is continued to use for general industrial facility, the following values for dose reference levels from residual contamination are proposed:

- For workers of specified industrial facility – 1 mSv per year (0.9 mSv per year owing to work at the facility and 0.1 mSv per year owing to living within the supervised area)
- For the public living within the supervised area – 0.1 mSv per year

At the stage of planning works and radiation protection measures it's necessary to make final decision regarding values of specified normative levels involving stakeholders in decision making according to new ICRP Recommendations [11, 12].

4.4 Numeric Values of Derived Ecological Remediation Criteria

Factors of radionuclide environmental contamination should be used as derived criteria of ecological remediation. In case of the Gremikha TSF area where residual radionuclide contamination of soil defines situation of existing exposure of workers, radionuclide concentration in soil can be such factor. Table 3 contains the values of derived criteria – radionuclide concentrations in surface soil layer when annual dose of external exposure of workers due to the residual contamination doesn't exceed 1 mSv per year that corresponds to the dose reference levels due to residual contamination proposed earlier. These values were derived using out door occupancy period of 2,000 hours per year. This period is too big for real environmental conditions at Far North, and dose of actual exposure at levels of contamination given in Table 3 will be much less than 1 mSv per year.

Table 3 Derived criteria of ecological remediation

Radionuclide[a]	Dose conversion factor (μSv/h)/(Bq/g)	DCER[b] (Bq/g)	Reference
Co-60	0.47	1	[8]
Sr-90+	0.003	200	[36]
Cs-137+	0.10	5	[8]
Eu-152	0.20	3	[8]
Eu-154	0.22	2	[8]
Summa of Pu-239, Pu-238 and Am-241	0.82	0.5	[8]

[a] "+" means that specified radionuclide is in equilibrium with daughter radionuclides
[b] Exposure during 2,000 hours per year is presumed

5 Scenario of the Environmental Remediation of Gremikha TSF

According to the developed strategy, works on the ecological remediation of Gremikha TSF are divided into four stages:

Stage 1. (2005–2010) – conversion of the Gremikha TSF to bring it into correspondence with current regulatory requirements

Stage 2. (2008–2015) – temporary operation of Gremikha TSF to remove SNF and RW from the site

Stage 3. (2010–2020) – decommissioning of the Gremikha TSF with removal of secondary RW generated at it

Stage 4. (2018–….) – remediation of the former Gremikha TSF site and adjacent areas

Each stage of the ecological remediation of Gremikha TSF should be characterized by the end-state that corresponds to the certain level of hazard of this radiation hazardous facility and can be fixed for a long time. Degree of the facility radiation hazard for workers, the public and the environment successively reduces during its environmental remediation.

General scenario of the ecological remediation of Gremikha TSF is a composition of scenarios for work performance provided at each stage of environmental remediation. Scenario of works at each stage includes initial and end state of the ecological remediation subject as well as sequence of work performance at considered stage. As successive stages are related by end and initial states while amount, composition and conditions of works at present can be estimated with significant uncertainty, scenarios for stages of the ecological remediation of Gremikha TSF should be refined in the course of works. Scenario for the stage of ecological remediation is a base for development of work engineering design.

During planning works it should be considered the fact that uncertainty of scenario for implementation of each stage increases as time period between development of scenario and start of planned work increases.

Analysis of radiometric survey results showed that contaminated areas where soil falls in the category of RW according to SPORO-2002 [8] are (see Fig. 1):

1. Surroundings of the Open Pad
2. An area under the rock near building 1B
3. Perimeter of building 1A from the side of the rock
4. LRW storage area

A scenario of the decommissioning and remediation of contaminated areas is considerably defined by amount of secondary RW that should be removed from the site. To determine radioactive waste that is given by actual SPORO-2002 [8] only the value of specific activity is used. At that, numeric values of criteria for solid radioactive waste are taken from NRB-99 [31] where ability to use such criteria is resulting from the value of released activity. At present IAEA issued a document included in series of IAEA Safety Standards No RS-G-1.7 [36] where derived release levels applied to "unlimited" amount of radioactive substance were defined. Such levels are used to determine a lower limit of specific activity for materials that should be referred to very low-level radioactive waste (VLLW). For example, this inconsistence leads to the fact (see Table 4) that solid waste with specific activity ^{90}Sr equal to 50 Bq/g which, according to SPORO-2002, are not radioactive, according to IAEA Standards [33] can be considered as radioactive waste where ^{90}Sr concentration is 50 times more than release level. That IAEA Standard is planned to be introduced in Russia in the nearest 5 years.

Figure 1 shows consequences of establishing a VLLW category for Gremikha case. According to SPORO-2002 [33] wastes with Cs-137 concentration $1 \times 10^7 - 1 \times 10^3$ Bq/g are treated as medium level waste, with Cs-137 concentration $1 \times 10^3 - 10$ Bq/g – as low level waste and material with concentration below 10 Bq/g could be cleared. This classification differs from recommended by the IAEA [5]. If category of VLLW will be introduced, it may include materials with Cs-137 concentration of 10–0.1 Bq/g [7]. Volume of such material will be huge in comparison with LLW.

Table 4 Levels of exception, elimination and clearance of radioactive materials from regulatory scope

	Specific activity (Bq/g)		
	Liquid and solid radioactive waste	Liquid radioactive waste	Solid radioactive waste
Radionuclide	BSS-2004 [7][a]	SPORO-2000 [33]	BSS-1996 [6][b] NRB-1999 [31][b] SPORO-2000 [33]
H-3	100	77	1,000,000
C-14	1	2.4	10,000
Co-60	0.1	0.41	10
Sr-90	1	0.05	100
Cs-137	0.1	0.11	10
Eu-152	0.1	0.99	10
Eu-154	0.1	0.69	10
Pu-239	0.1	0.0006	1

[a] Unlimited amount of cleared material
[b] Limited amount of cleared material

Decrease of limit for reference of materials containing radionuclides to radioactive materials is inevitable and, first of all, will be aimed to toughening requirements to control over management of man-caused radiation sources. It completely corresponds to the concept of radiation source management that is a base of the public radiation safety assurance. In case of the environmental remediation of Gremikha TSF, contaminated soil is not a consequence of regulated practical activity for which anyone can be responsible because the Gremikha TSF as a whole is radiation legacy of the former USSR. In this case toughening regulatory requirements to soil residual contamination that leads to situation of existing exposure will not cause increase of safety of this contamination source because this source hasn't existed anymore. So, to solve general issue of Radiation Legacy Facility remediation it's necessary to introduce Recommendations of ICRP [2, 10, 12, 13] and IAEA [9] into national normative basis of radiation safety regulation. According to the ICRP Recommendations [12], situation of existing exposure should be regulated as at present exposure induced by natural radiation sources (in particular – radon) is regulated. At such approach environments contained radionuclides (soil, bottom sediments) are not considered as radioactive waste required to be removed, if annual effective dose in situation of existing exposure doesn't exceed specified level – for example, levels proposed in previous section. Comparison of values of derived criteria for the ecological remediation of Gremikha TSF site presented in Table 3 with release levels presented in Table 4 conforms expediency of introduction of international recommendations in national practice.

STAGE 1. CONVERSION OF THE GREMIKHA TSF

The main content of work scenario at the Gremikha TSF conversion stage is performing activities for putting its condition into compliance with active regulative requirements of nuclear and radiation safety assurance to the SNF and RW temporary storage. Objectives at this stage are:

- Comprehensive engineering and radiation examination
- Inventory taking and providing proper storage of SNF and RW
- Providing safe conditions for work performance
- Development a strategy of the ecological remediation of Gremikha TSF and scenarios for its implementation of individual stages
- Maintenance of nuclear and radiation safety at the Gremikha TSF

Because the Gremikha TSF is a Radiation Legacy Facility of the former USSR, input data about its condition is incomplete. In such conditions works at the stage 1 are planned and performed at higher radiation hazard. So the priority work direction at this stage is radiation safety assurance for workers.

Potential sources of the personnel exposure at Gremikha TSF are:

- Technical facilities
- The area contaminated with man-caused radionuclides (some spots of the area)
- Surface radioactive contamination of process equipment and surfaces of working rooms
- Contaminated with man-caused radionuclides transport means and process equipment forwarded to repair and to their temporary storages

- Processes of the Gremikha TSF renovation, conservation and liquidation leading to intensive intake of radionuclides to working rooms and environments
- Radioactive waste generated during remediation of environments

Hygienic base of radiation safety assurance and environmental protection is zoning – partition of the facility rooms and area into radiation hygienic zones based on the level of radiation hazard, with setting appropriate limits for radiation factors.

Zoning of the Gremikha TSF area in conditions of temporary operation should conform to requirements of MU 2.6.6.22-05 [30]. There are the following radiation hygienic zones established at the TSF area:

- Controlled access zone (CAZ)
- Free access zone (FAZ)

In the CAZ only the personnel from A group is allowed to work owing to personal monitoring of external and internal exposure.

About 20% of the CAZ area has soil contamination with ^{137}Cs over 1 Bq/g. At that, these contaminations are located mainly near four objects: the Open Pad, LRW storage area, building 1 and at the area under the rock as shown at Fig. 1.

Personnel radiation monitoring should conform to the regulations approved in the proper order and be performed according to requirements of MU 2.6.1.16-2000 [22]. At that:

1. Monitoring of external exposure – according to MU 2.6.1.25-2000 [23]
2. Monitoring of internal exposure – according to MU 2.6.1.26-2000 [24], MUK 2.6.1.09-03 [29] and MVR 2.6.1.50-01 [26]
3. Monitoring of radiation situation – according to MU 2.6.1.14-01 [25] and MU 2.6.1.44-02 [27]
4. Monitoring of exposure of skin and lens – according to MU 2.6.1.56-02 [28]

STAGE 2. TEMPORARY OPERATION OF THE GREMIKHA TSF

The main content of work scenario at the stage of the temporary operation of Gremikha TSF is removal of SNF and primary storage RW from the site. Objectives at this stage are:

- Examination, inventory taking and providing proper storage of SNF and RW
- SNF and RW removal from the site
- Decontamination of the facility buildings and structures
- Providing safe conditions for work performance and
- Determination of the end-state of the Gremikha TSF site and adjacent areas after ecological remediation completion

STAGE 3. DECOMMISSIONING OF THE GREMIKHA TSF

The main content of work scenario at the stage of Gremikha TSF decommissioning is partial or full dismantling equipment, buildings and structures as well as making final decision regarding end-state of the Gremikha TSF site and adjacent areas after ecological remediation completion. The plan of the works is based on final decision regarding the end-state of the Gremikha TSF site after ecological remediation completion.

The stage is closed with full completion of works on dismantling contaminated building and structures of Gremikha TSF.

STAGE 4. REMEDIATION OF THE FORMER GREMIKHA TSF SITE AND ADJACENT AREAS
The main content of work scenario at this stage is remediation of the former Gremikha TSF site, adjacent areas and water area within the Sanitary buffer zone according to a certain end-state. The plan of the works is based on final decision regarding the end-state of the Gremikha TSF site after ecological remediation completion.

The stage is closed with full completion of works on the ecological remediation of Gremikha TSF after comprehensive examination that confirms reaching the area residual contamination level corresponding to certain end-state of the Gremikha TSF site after ecological remediation completion.

The Gremikha TSF operation has the main impact on the area within Sanitary buffer zone, firstly within the controlled area (industrial site) and on the coastal water area. At this area there are local spots with significant soil contamination with man-caused radionuclides and with higher dose rate levels formed by both soil contamination with radionuclides and radiation due to radioactive materials located in buildings.

The results of radiation monitoring and radiation situation examination in the area of the Gremikha TSF location serve as an evidence that it is impossible to find man-made impact of the Gremikha TSF operation on the territory of Surveillance zone and the public living within it, at the level of natural background and global precipitations. The forthcoming operations for ecological remediation of Gremikha TSF will be conducted in accordance to contemporary regulation, thus its environmental impact will be negligible.

6 Conclusions and Recommendations

Multifactor analysis of strategies for three Gremikha TSF representative facilities showed that the most acceptable is strategy of stage-by-stage environmental remediation. Process of ecological remediation of Gremikha TSF should be divided into completed intermediate stages characterized by successive increase of the whole facility protection and safety. At each of these stages the Gremikha TSF as a whole and each individual facility should be manageable, stable, protected and safe.

The strategy of the stage-by-stage remediation of Gremikha TSF includes:

At present. Conversion of the Gremikha TSF facilities to use for their initial purpose

In the years coming. In the course of preliminary works:

- Preparation of protected and safe places for interim storage of SNF and primary RW at the Gremikha TSF site
- Moving of SNF and primary RW from existing storages to temporary ones and then – to reprocessing plants of repositories at "continent"

- Decommissioning some of the Gremikha TSF facilities and their function change to new radiation hazardous facilities – interim storages for secondary RW

During 10–15 years. Possible long-term storage of secondary low-level RW and structures (waste) with fixed contamination at the Gremikha TSF site until making a decision regarding the end-state of the Gremikha TSF site.

In far prospect. Transformation of the RW temporary storages at the Gremikha TSF site to disposal places or their liquidation with cleanup of the Gremikha TSF site and coastal water area up to appropriate level resulting from decision regarding further use (end-state) of the Gremikha TSF.

Thus, during no less than 12–15 years the Gremikha TSF will be in the regime of operation of the radiation hazardous facility functioning under the active legislative area and staying under regulatory control. At intermediate stages all Gremikha TSF facilities should meet these requirements.

A need of supplementary normative and legislative support becomes actual at the later stage of works on ecological remediation in case of works implemented for RW disposal or any options of further non-radiation technological use of the Gremikha TSF area. Solution of this issue is promoted by introduction in national practice of new ICRP Recommendations and IAEA radiation safety assurance norms at management of radiation sources. This, in its turn, determines approach to statement and solution sequence for the task of development of criteria for final ecological remediation of the Gremikha TSF, including:

1. Development of recommendations and solutions for use of active norms and laws during preliminary period.
2. Development and putting in force of supplementary normative and legislative regulations by the time the period of using facilities with changed function is expired.

Using multifactor analysis of expert judgments during selection of acceptable strategy for the ecological remediation of Gremikha TSF showed high efficiency. This approach allowed analyzing a wide range of possible end-states of the facility after its remediation and substantiating the most acceptable strategy of ecological remediation in short terms and with minimum costs. Methodology of multifactor analysis used to estimate options of strategy for the ecological remediation of Gremikha TSF can be extended for other NRHF too. At the beginning of this work it's necessary:

- To determine a range of possible end-states of the facility after its remediation
- To develop options of the ER strategy
- To determine stakeholders

Important stage of this work is selection of experts that should provide involvement of specialists of such amount and qualification level that they fill all "niches" in discussed issue of considered facility environmental remediation. After that it's necessary, with participation of all experts, to list factors based on which strategies of ecological remediation will be estimated.

Acknowledgments This study has been supported by the French Atomic Energy Commission under Contract No. 4000264028 from 22.12.2006.

List of Abbreviations

CATF	closed administrative-territorial formation
CAZ	controlled access zone
ER	environmental remediation
FAZ	free access zone
Gremikha TSF	facility for spent nuclear fuel and radiation waste temporary storage in Gremikha
LLW	low-level radioactive waste
LRW	liquid radioactive waste
NRHF	nuclear and radiation hazardous facility
RLF	Radiation Legacy Facility
RW	radioactive waste
SBZ	sanitary buffer zone
SevRAO	Federal state unitary enterprise for radioactive waste management in the Northwest region
SNF	spent nuclear fuel
SRW	solid radioactive waste
TSF	facility for spent nuclear fuel and radiation waste temporary storage
VLLW	very low-level radioactive waste

References

1. Bylkin B K, Gorlinsky Yu E, Kutkov V A, Nikolsky O A, Pavlenko V I, Sivintsev Yu V, Stepennov B S (2007) Use of multifactor analysis for scoring final states and strategies for ecological rehabilitation of the Facility for temporary storage of spent nuclear fuel and radioactive waste in Gremikha. Preprint IAE 6456/3. RRC "Kurchatov Institute", Moscow (in Russian).
2. Gonsales A J (2001) Decision making in view of chronic public exposure. New ICRP Recommendations. ANRI 3(26):37–49 (in Russian).
3. Gorlinsky Yu E, Nikolsky O A, Papkovsky B P, Sivintsev Yu V, Stepennov B S, et al. (2005) STB "Gremikha": Conditions, problems and perspectives of the environmental remediation. Izv. RAN. Energetika 5:28–37 (in Russian).
4. Gorlinsky Yu E, Nikolsky O A, Sivintsev Yu V, Stepennov B S, et al. (2005) Challenge of the ecological remediation of the former STB of Russian Navy in Gremikha village in terms of development predictions and tendencies of CATF Ostrovnoj of Murmansk region. Bulletin of Issues of Dismantling Nuclear Submarines 4(8):14–19 (in Russian).
5. IAEA (1994) Classification of radioactive waste. Safety Series No. F-111-G-1.1. IAEA, Vienna.
6. IAEA (1996) International basic safety standards for protection against ionizing radiation and for the safety of radiation sources. Safety Series No. 115. IAEA, Vienna.

7. IAEA (2004) Application of the concepts of exclusion, exemption and clearance. Safety guide. Safety Standards Series No. RS-G-1.7. IAEA, Vienna.
8. IAEA (2005) Derivation of activity concentration values for exclusion, exemption and clearance. Safety Reports Series No. 44. IAEA, Vienna.
9. IAEA (2006) Release of sites from regulatory control on termination of practices. Safety guide. Safety Standards Series No. WS-G-5.1. IAEA, Vienna.
10. ICRP (1999) Protection of the public in situations of prolonged radiation exposure. ICRP Publication 82. Annals of the ICRP 29(1–2). Pergamon Press, Oxford.
11. ICRP (2006) Assessing dose of the representative person for the purpose of radiation protection of the public and the optimization of radiological protection: Broadening the process. ICRP Publication 101. Annals of the ICRP 36(3). Pergamon Press, Oxford.
12. ICRP (2007) Recommendations of the ICRP. ICRP Publication 103. Annals of the ICRP 37(2–3). Pergamon Press, Oxford.
13. Kutkov V A (2007) Evolution of the system of radiation safety assurance in the light of new ICRP and IAEA Recommendations. ANRI 1(48):2–24 (in Russian).
14. Kutkov V A, Bezrukov B A, Tkachenko V V, Romantsov V P, et al. (2002) General provisions and requirements of normative documents in practice of assurance of radiation safety at nuclear power plants. Kutkov V A and Bezrukov B A (Eds.). Concern "Rosenergoatom", Moscow (in Russian).
15. Kutkov V, Kochetkov O, Panfilov A (2002) Strategy of control at source as a base for protecting workers against risks arising from exposure to ionizing radiation in the Russian Federation – in Occupational radiation protection: Protecting workers against exposure to ionizing radiation (Contributed Papers at the International Conference of IAEA, 26–30 August 2002), 39–44. IAEA, Vienna.
16. Kutkov V A, Tkachenko V V, Romantsov V P (2003) Radiation safety of the personnel of nuclear power plants. Kutkov V A (Ed.). Atomtechenergo. IATE, Moscow-Obninsk (in Russian).
17. Marey A N, Barkhudarov R M, Novikova N A (1974) Global precipitations of cesium-137 and human being. Atomizdat, Moscow (in Russian).
18. Moiseev A A (1985) Cesium-137, environment, human being. Energoatomizdat, Moscow (in Russian).
19. R.F. Federal Law No. 1244-1 (1991) About social protection of the citizens affected by radiation due to accident at the Chernobyl NPP (with modifications of editions of Federal Laws NN. 3061-1 (1992), 179-FZ (1995), 149-FZ (1996), 44-FZ (1997), 79-FZ (1999), 127-FZ (1999), 122-FZ (2000), 5-FZ (2001), 110-FZ (2001), and 189-FZ (2001)) (in Russian).
20. R.F. Federal Law No. 149-FZ (1995) About social protection of the citizens affected by radiation due to nuclear tests at the Semipalatinsk test site (in Russian).
21. R.F. Federal Law No. 175-FZ (1998) About social protection of citizens of the Russian Federation affected by radiation due to the accident in 1957 on a production association "Mayak" and dumps of radioactive waste products into the Techa River (in Russian).
22. R.F. MEDBIOEXTREM (2000) Determination of individual effective and equivalent doses and organization of occupational exposure monitoring at controlled conditions of radiation source management. General requirements. Methodical guidance MU 2.6.1.16-2000. Kutkov V A, Jaryna V P, Popov V I, et al. Russian Federal Agency of Medical Biological and Extreme Issues and Ministry of the Russian Federation for Atomic Energy, Moscow – in Methodical support of radiation monitoring at the enterprise 1(2001):23–55. Ministry of the Russian Federation for Atomic Energy, Russian Federal Agency of Medical Biological and Extreme Issues, Moscow (in Russian).
23. R.F. MEDBIOEXTREM (2000) Radiation monitoring of external occupational exposure. General requirements. Methodical guidance MU 2.6.1.25-2000. Polenov B V, Mysev I P, Petrov V I, Sokolov A D, Kutkov V A., et al. Russian Federal Agency of Medical Biological and Extreme Issues and Ministry of the Russian Federation for Atomic Energy, Moscow – in Methodical support of radiation monitoring at the enterprise 1(2001):57–110. Ministry of the Russian Federation for Atomic Energy, Russian Federal Agency of Medical Biological and Extreme Issues, Moscow (in Russian).

24. R.F. MEDBIOEXTREM (2000) Radiation monitoring of internal occupational exposure. General requirements. Methodical guidance MU 2.6.1.26-2000. Kochetkov O A, Popov V I, Kutkov V A, et al. Russian Federal Agency of Medical Biological and Extreme Issues and Ministry of the Russian Federation for Atomic Energy, Moscow – in Methodical support of radiation monitoring at the enterprise 1(2001):111–155. Ministry of the Russian Federation for Atomic Energy, Russian Federal Agency of Medical Biological and Extreme Issues, Moscow (in Russian).
25. R.F. MEDBIOEXTREM (2001) Monitoring of radiation situation. General requirements. Methodical guidance MU 2.6.1.14-01. Kovalenko V V, Artemenkova L V, Mysev I P, Kutkov V A, et al. Russian Federal Agency of Medical Biological and Extreme Issues and Ministry of the Russian Federation for Atomic Energy, Moscow – in Methodical support of radiation monitoring at the enterprise 1(2001):157–183. Ministry of the Russian Federation for Atomic Energy, Russian Federal Agency of Medical Biological and Extreme Issues, Moscow (in Russian).
26. R.F. MEDBIOEXTREM (2001) Determination of radionuclide intake and individual effective dose based on results of measurements on WBC of radionuclide concentration in human body for workers of nuclear power plants. Procedure of conducting calculations MVR 2.6.1.50-01. Kutkov V A, Tkachenko V V, Romantsov V P. Russian Federal Agency of Medical Biological and Extreme Issues and Concern "Rosenergoatom", Moscow – in Methodical support of radiation monitoring at the enterprise 5(2005):75–99. Ministry of the Russian Federation for Atomic Energy, Russian Federal Agency of Medical Biological and Extreme Issues, Moscow (in Russian).
27. R.F. MEDBIOEXTREM (2002) Radionuclide activity concentration in the air at workplaces. Requirements to determination of average annual activity. Methodical guidance MU 2.6.1.44-02. Usoltsev V J, Kutkov V A, Yeltsin V F, et al. Russian Federal Agency of Medical Biological and Extreme Issues and Ministry of the Russian Federation for Atomic Energy, Moscow – in Methodical support of radiation monitoring at the enterprise 5(2005):57–74. Ministry of the Russian Federation for Atomic Energy, Russian Federal Agency of Medical Biological and Extreme Issues, Moscow (in Russian).
28. R.F. MEDBIOEXTREM (2002) Monitoring of equivalent doses of photon and beta radiation in skin and lens. Methodical guidance MU 2.6.1.56-02. Shaks A I, Gimadova T I, Timofeev L V. Russian Federal Agency of Medical Biological and Extreme Issues and Ministry of the Russian Federation for Atomic Energy, Moscow (in Russian).
29. R.F. MEDBIOEXTREM (2003) Regulations of radiation monitoring of internal exposure of nuclear power plant workers. General requirements. Methodical guidance for radiation monitoring MUK 2.6.1.09-03. Kutkov V A, Tkachenko V V, Romantsov V P. Russian Federal Agency of Medical Biological and Extreme Issues and Concern "Rosenergoatom" of the Russian Federation for Atomic Energy, Moscow – in Methodical support of radiation monitoring at the enterprise 5(2005):41–56. Ministry of the Russian Federation for Atomic Energy, Russian Federal Agency of Medical Biological and Extreme Issues, Moscow (in Russian).
30. R.F. Medical-biological agency (2005) Radiation safety assurance at works on remediation of coastal technical bases areas. Methodical guidance MU 2.6.6.22-05. Russian Federal medical-biological agency, Moscow (in Russian).
31. R.F. MPH (1999) Radiation safety norms (NRB-99): Hygienic regulations SP-2.6.1.758-99. Russian Ministry of Public Health, Moscow (in Russian).
32. R.F. MPH (1999) Basic sanitary regulations of radiation safety assurance (OSPORB-99). Sanitary regulations SP-2.6.1.799-99. Russian Ministry of Public Health, Moscow (in Russian).
33. R.F. MPH (2002) Sanitary regulations of radioactive waste management (SPORO-2002): Sanitary regulations SP-2.6.6.1168-02. Russian Ministry of Public Health, Moscow (in Russian).
34. Shandala N K, Kiselev M F, Sneve M K, et al. (2006) Radiation environmental normalization in conditions of remediation operations at "SevRAO". – in Current issues of the public radiation safety assurance (Composite book of reports and theses of Scientific and practical conference,

St-Petersburg, 4–7 December 2006), 184–186. Federal Agency on supervision in sphere of protection of consumer rights and human well-being, St-Petersburg (in Russian).
35. Sivintsev Yu V, Vysotsky V L, Kalinin R I, et al. (2006) Quantitative criteria of the site remediation of shore technical bases. Atomnaya Energiya 101(1):35–49.
36. U.S. EPA (2001) Cancer risk coefficients for environmental exposure to radionuclides: CD supplement. Federal Guidance Report 13. EPA 402-C-99-001. U.S. Environmental Protection Agency, Washington, DC.
37. UNSCEAR (2000) Report to the General Assembly. Volume 1: Sources, Annex B: Exposures from natural radiation sources. United Nations, New York.

Transport-Technological Scheme of High-Level SRW Management from the Reactor Facilities in the Northwest Russia

V.A. Mazokin

Abstract This paper describes the transport-technological scheme of high-level solid radioactive waste management from the reactor facilities in the Northwest Russia, taking into account the regulatory requirements as applied to different waste categories.

Keywords *SPORO-2002*, SRW, «dry» residues, decommissioning

The Sanitary Rules for the Radioactive Waste Management *(SPORO-2002)* SP 2.6.6. 1168–02, as one of the RF regulative documents, establishes the classification of solid radioactive wastes (SRW). According to this classification the SRW category is defined in terms of their specific activity, radioactive contamination level and gamma dose rate at 0.1 m distance from the surface (see Table 1).

1 Solid Radioactive Waste Categories

The primary detection of the SRW category is implemented by means of beta- and gamma-level measurements (column **VI** and **IX**).

High level SRW category includes equipment, constructions, and materials, radioactive beta-level of which (as well as alpha and Transuranium radionuclides) is higher than $1 \cdot 10^7$ part/cm^2 min, or gamma dose rate at 0.1 m distance from their surface is higher than 10 mGy/h.

Radioactive contamination is generally superficial, so its level can be reduced by means of decontamination. SRW with such kind of contamination may be transferred into the lower category and they are beyond the scope of this paper.

High gamma radiation level specifies the radioactivity concentration, i.e., the material structure includes radionuclides; therefore, decontamination methods can only reduce beta radiation contribution into the total spectrum.

FSUE «NIKIET named after NA Dollezhal», Russia

M.K. Sneve, M.F. Kiselev (eds.) *Challenges in Radiation Protection and Nuclear Safety Regulation of the Nuclear Legacy*,

Table 1 Solid radioactive waste categories

No	Waste category	Specific activity (kBq/kg)			Radioactive contamination level (part/cm² min)			Gamma dose rate (in 0.1 m from the surface) (mGy/h)
		Beta radiation of the radionuclide	Alpha radiation of the radionuclide	Transuranium radionuclide	Beta radiation of the radionuclide	alpha radiation of the radionuclide	Transuranium radionuclide	
1	Low level	Less than 1E3	Less than 1E2	Less than 1E1	5 × 1E2 ÷ 1E4	5 × 1E2 ÷ 1E4	5 × 1E2 ÷ 1E4	0.001 ÷ 0.3
2	Intermediate level	1E3 ÷ 1E7	1E2 ÷ 1E6	1E1 ÷ 1E5	1E4 ÷ 1E7	1E3 ÷ 1E6	1E2 ÷ 1E5	0.3 ÷ 10
3	High level	More than 1E7	More than 1E6	More than 1E5	More than 1E7	More than 1E6	More than 1E5	More than 10

High level SRW (according to their gamma dose rate) resulted from operation and decontamination of ship reactors include first of all reactor internals being removed from the reactors in the course of their maintenance and re-fuelling (units of the reactor power and containment regulation, removable constructions etc.).

Filtering substances (including the batch mixture of the coolant cleaning filters of the primary and third coolant circuits), operation life of which has been terminated, being removed from the filter vessels, are also considered as high level SRW. They are conveyed to the shore storage facilities (the filter batch mixture consists of such radionuclides as **Cs-137, Ba-139, Cr-51, Co-60** etc.).

Special protective containers (filter retainer (catch tank)) are applied to accept and store the spent batch mixture; these containers are transported to the shore storage facilities.

«Dry» residues (concentrate) of the LRW processing and bottom sediments in the SNF storage facilities can also be high level radioactive.

Table 2 includes some summarized data on high level SRW from the ship reactors.

Now, storage of cartridges with the CPS absorber control rods is being implemented:

Table 2 List of high level SRW

No	SRW type	Current mode of storage	Place of storage	Amount, *units.*	Option of the final isolation
	Absorber control rod cartridges of the nuclear submarine reactors	(1) In SNF containers; (2) In CPS storage facilities in the Floating Engineering Tanks (FET) (without containers)	Shore storage facilities in Andreeva, Gremikha, in FET storage facilities	About 5,500	In reactors of the decommissed nuclear submarines
	Sorbets (batch mixtures) of activity filters from the primary and third coolant circuits of the nuclear submarine reactors	In special catch tanks	RW shore storage facilities	330 (in Andreeva – 306, In Gremikha – 22)	In containers, in the regional centre for RW storage
	Hardened concentrates resulted from LRW processing (purification)	In SRW storage facilities	LRW processing facilities	Undetermined	In containers, in the regional centre for RW storage
	Residues (bottom sediments) in SNF and RW storage facilities	Package-free	In storage tanks in FET and in the shore RW storage facilities	Undetermined	Hardening and isolation in the available construction

(continued)

Table 2 List of high level SRW

No	SRW type	Current mode of storage	Place of storage	Amount, *units.*	Option of the final isolation
	Ionization chamber assemblies	In containers	In storage facilities in FET and in the shore RW storage facilities	133 (in Andreeva)	Undetermined

Note: Type and amount of high level SRW are to be ascertained according to the examination results of the RW storage facility, as well as on the base of the adopted technologies of their processing

- In the floating engineering tanks in the CPS storage facilities
- In the shore storage facilities in the ChT-type containers intended for Irradiated Fuel Assemblies (IFA) (see Fig. 1)

Catch tanks for the activity filter sorbets are concrete containers of two types:

- 901 drums (Ø 360 mm, height 1,700 mm)
- 3801 drums (Ø 520 mm, height 2,500 mm)

Figure 2 shows the allocation of the catch tanks in the storage facility in Andreeva Bay.

IC assemblies, emplaced in the storage facilities in Andreeva Bay, are of the following dimensions: Ø 110 mm, length 1,550 mm.

The above-mentioned SRW are being stored in the interim packages and in the temporary storage facilities, which do not meet to the up-to-date requirements of the current regulative documents on radiation and environmental safety assurance with respect to such products and buildings.

Now, works begin aimed at Navy ship decommissioning and dismantlement as well as at the ecological remediation of shore radiation hazardous sites, where RW is being stored.

Programs of these works envisage conditioning, strong package, removal from the site and long-term isolation (disposal) of the SRW. In this light, the selection and justification of the SRW management scheme, especially high level SRW, become very important.

The final isolation (disposal) of the CPS cartridges is the most relevant.

FSUE «NIKIET» has developed and introduced the technology of the final isolation of the reactor CPS cartridges during NS decommissioning and dismantlement. This technology is based on the provision of the state concept of the overall NS decommissioning.

The reactor vessels, after the SNF removal, and equipment of the steam generating installation as a part of the reactor compartments are subjected to the long-term storage for decay (about 70–100 years) in the special shore storage facilities.

Duration of such storage is sufficient to reduce radioactivity of the metal containing **Co-60**, **Fe-55**, and **Ni-59** nuclides up to the level allowing its re-use.

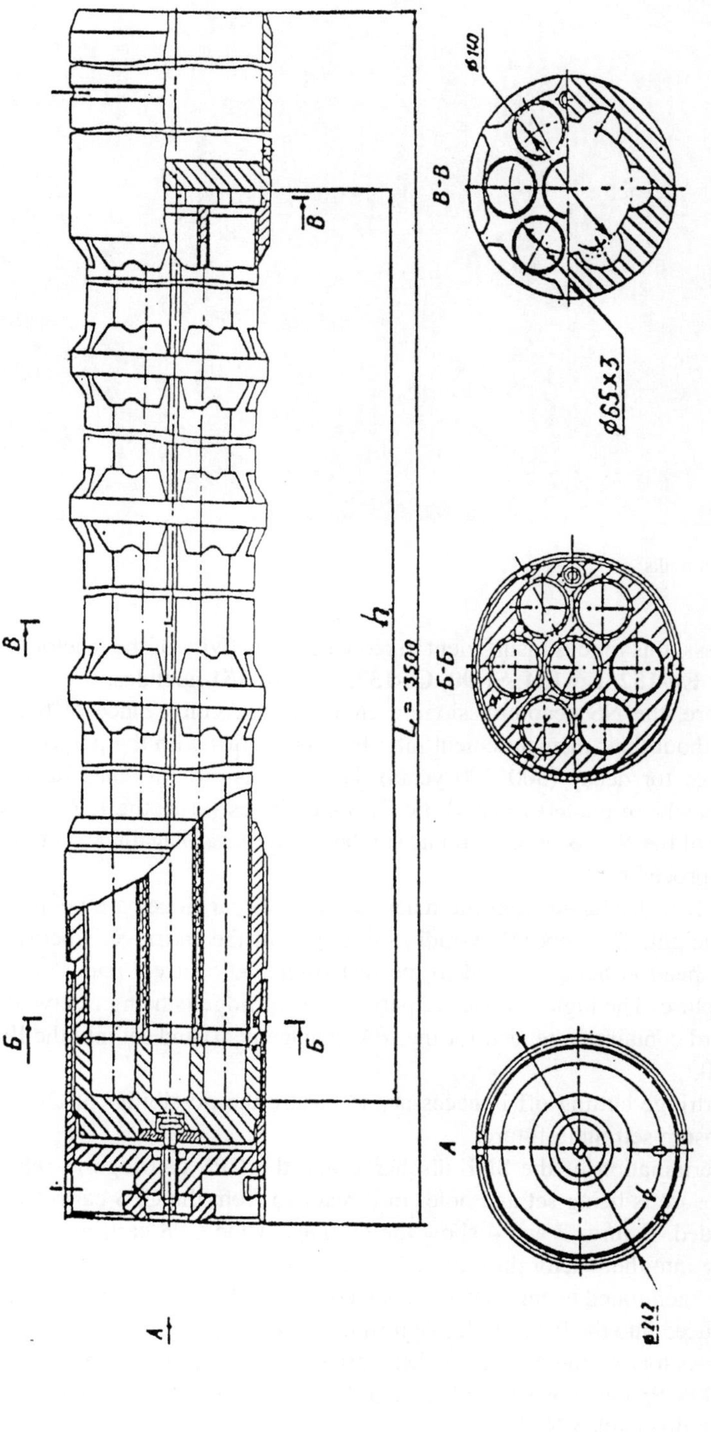

Fig. 1 ChT-type container

Fig. 2 Catch tanks

Nevertheless, this time is insufficient to reduce radioactivity of the reactor internals containing **Eu-152**, **Eu-154**, **Sr-90**, **Cs-137**, and **Sm-151** nuclides.

Therefore, the NS decommissioning envisages the conveyance of the reactor vessels without their dismantlement into the repositories with the purpose of their final storage for decay (300–500 years). Having this circumstance in mind, the decision has been made to install the CPS cartridges (after the IFA removal in the course of the NS decommissioning) in the reactor vessel again, according to the following procedure.

During IFA discharge from the reactors, the CPS cartridges are being lifted at the fixed height. The upper (low-radioactive) part of the cartridge is being cutting off, and a head is being welded to the rest (high-radioactive) part to ensure its further capture. The high-radioactive part of the cartridge is being removed (using the standard container intended for the IFA management) and set into the IFA-free reactor cell.

The cartridge cutting off is necessary to ensure the possibility of the standard reactor closure setting in future.

After termination of the SNF discharge and the CPS cartridge re-setting, the reactor closure is being set and holes in it resulted from the CPS cartridge set are being welded. Figures 3 and 4 show the outlines of the CPS cartridge cutting off and setting into the reactor during the NS decommissioning.

The above-mentioned technology has been coordinated with the supervision bodies and introduced into the list of technological operations relating to the SNF discharge from the reactors of the NS under dismantlement. According to this technology, about 1,600 CPS cartridges have been emplaced in the reactors located the reactor units of the dismantled NS.

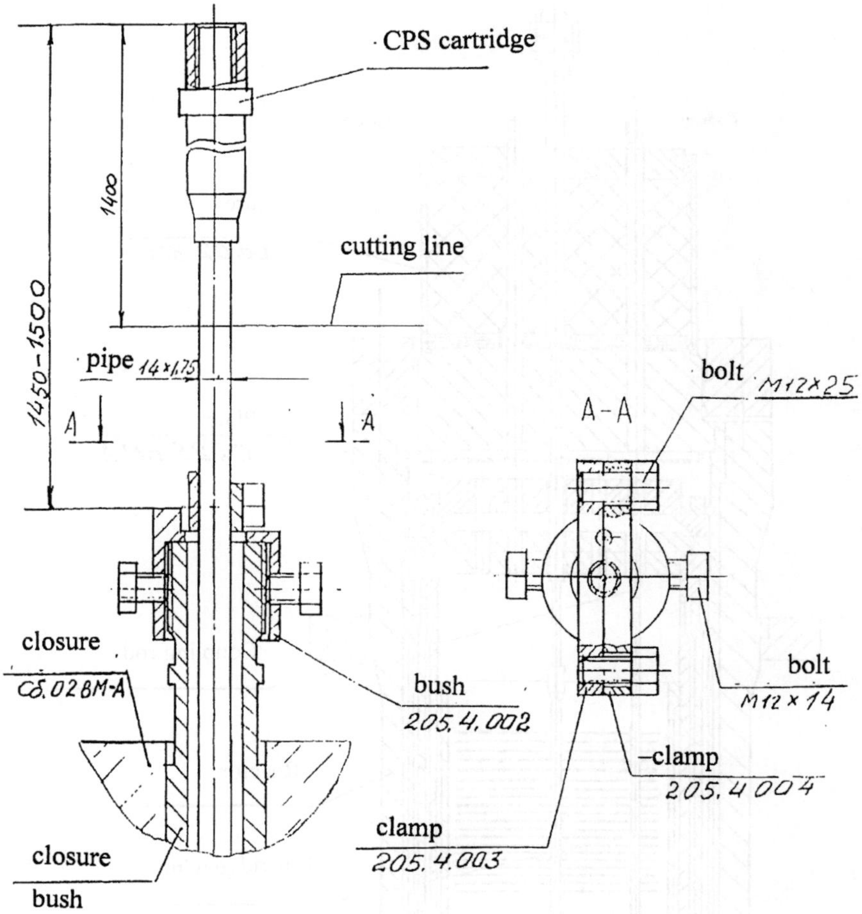

Fig. 3 Outline of the CPS cartridge cutting off

As for the problem of the CPS cartridge isolation (disposal), being removed in the course of the reactor re-charging and stored in the FET and in the shore storage facilities, this problem has not been being decided yet. The started design works connected with ecological remediation of the shore sites of RW storage, need urgent solution on the CPS cartridge management at this and other sites and facilities.

The most reasonable, from our point of view, is application of such CPS cartridge isolation, which is similar to that being applied during the NS decommissioning.

To implement it, the following engineering decisions and technologies are to be developed:

- To define locations, package type, storage conditions and amount of CPS cartridges needed their final isolation (disposal)
- To chose some (three to four) single units of the reactor compartments (RC) with two reactors to be assembled at "Nerpa" facility, and to prepare them for acceptance and CPS cartridge setting into their reactors

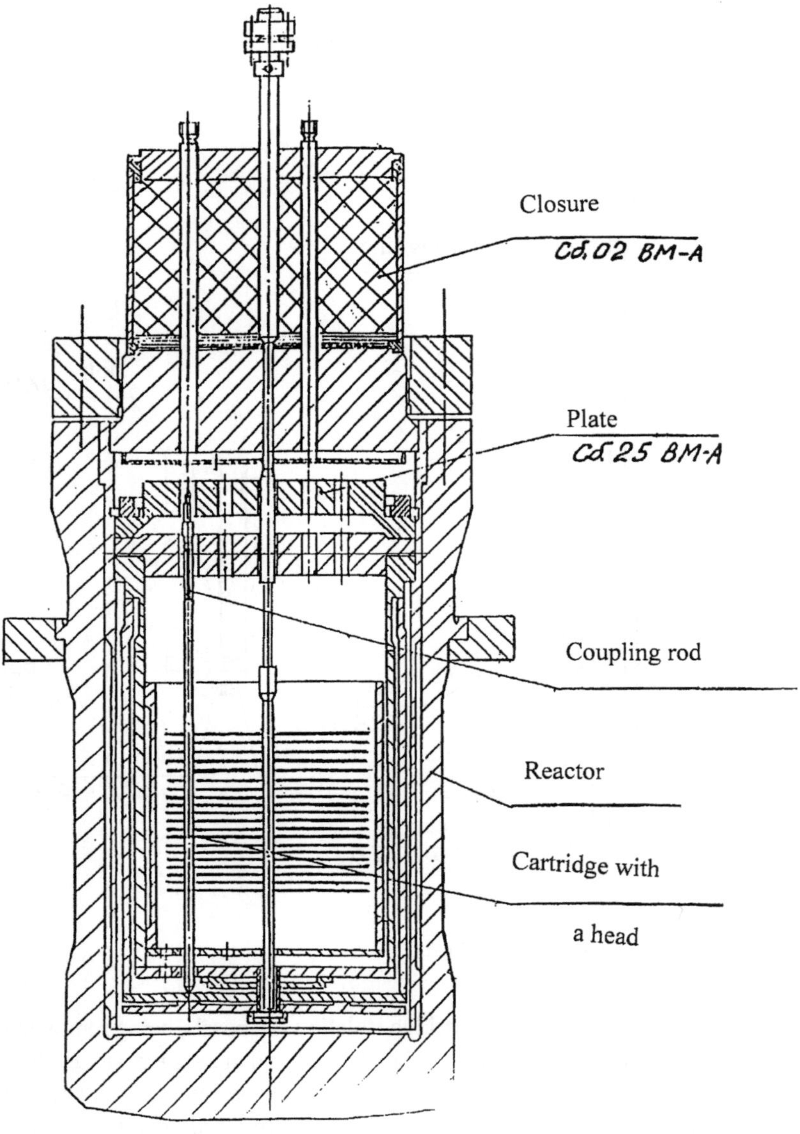

Fig. 4 Position of the cut CPS cartridges after their emplacement in the reactor cells

- To develop organization-engineering and technological documentation relating to the CPS cartridge preparation and conveyance from the place of their storage to "Nerpa" facility
- To develop the engineering documentation relating to the CPS cartridge setting into the reactors of the RC unit, which is being prepared for long-term storage in the CDL "Sajda"

- To emplace the CPS cartridges in the CR reactors, to complete RC assemblance and preparation for long-term storage according to the existing technology and to convey RC to the CDL "Sajda"

The implementation of the proposed technology will permit:

- To eliminate radiation hazardous operations and costs of conditioning, package, storage and following isolation (disposal) of high-level SRW, to reduce the space intended for their storage in the RW regional centre
- To solve the problem of isolation (disposal) of such high-level SRW as the CPS cartridges from the ship reactors, today

Management of other high-level SRW mentioned in Table 2, with the purpose of their final isolation, is assumed to be implemented according to the traditional technology: packaging into the containers (transport or dual-purpose: transportation and storage) and conveyance them to the regional centre for RW management.

2 Summary

1. Now, high-level SRW of the ship reactors to be subjected to final isolation (disposal) include:

    - Cartridges with absorber control rods of the ship reactor control and protection systems (CPS cartridges). Total amount of the spent CPS cartridges in the northwest region is about 7,000, and 1,600 of them are being emplaced again in the reactors with the purpose of final isolation; 206 are being emplaced in the storage facilities of the floating engineering tanks; 1,570 are being emplaced in the shore storage facilities (Andreeva Bay). Now, there is no decision on their final isolation.
    - Ionization chambers (IC) of the reactor monitoring system. There is 133 containers with IC in the storage facilities in Andreeva Bay. We have no data relating to other sites. During NS decommissioning, IC remain in the reactor compartment behind the standard containment barriers.
    - «catch tanks» of the activity filters of the primary and third coolant circuits. Total amount of the catch tanks is 328 (306 – in Andreeva Bay, 22 – in Gremikha village)
    - Hardened concentrates resulted from LEW purification and processing
    - Bottom sediments in tanks of IFA and LRW "wet" storage

2. Future management of high-level SRW is assumed to be implemented according to the following procedure:

    - To set cartridges with CPS control rods (which are being temporary stored in FET and in shore storage facilities) into cells of several IFA-free reactors. With this purpose, to chose three to four reactor compartments intended for allocation to store for decay in the CLS "Sajda"

- To convey «catch tanks» of filters, IC chambers, concentrates resulted from LRW processing to the regional RW management centre for their final isolation
- To harden bottom sediments and to isolate them in the tanks of their storage

The Techa Reservoir Cascade: Safety and Regulation Problems

Yu.G. Mokrov, Yu.V. Glagolenko, E.G. Drozhko, and S.I. Rovny

Abstract The paper provides a short historical overview of creation, usage and current radiation and hydrologic status of the Techa reservoir cascade. The analysis is given of the reservoir water balance, main water flows and sources of radionuclide filtration inflow from the TRC into the open hydrographic system of the Techa river.

It is demonstrated that the main problem of the TRC usage is related to the general tendency of water level growth in the reservoirs, which results in the increase of radionuclide inflow into the Techa river with filtration drains and forms additional hydrostatic load on the dam of the end cascade reservoir R-11.
Different options increasing safety of the TRC use currently implemented and planned for the future are analyzed.

Keywords Techa reservoir cascade, radioactive contamination, Muslyumovo settlement, TRC

1 Introduction

In 1949–1956, while implementing the State defense program, the "Mayak" PA performed routine (specified in the detail design) and accidental releases of liquid radioactive releases (LRW) into the Techa river. The LRW consisted of fission origin radionuclides with half-life from several days to dozens of years. As a result all the components of the Techa river (water, bottom sediments, floodplain, vegetation, biota) were subjected to large-scale radioactive contamination [1, 2, 4].

Before entering into the open hydrographic system of the Techa river, the LRW releases were first directed to the circulating water sedimentation reservoir R-3 (Koksharovsky reservoir) constructed in August, 1951 and then – to the reservoir R-4 (Metlinsky pond). In 1956 at the distance of 12 km below the LRW release point the riverbed was closed by the earth dam and the reservoir R-10 was created, which enabled complete termination of radionuclide inflow into the open hydrographic

Mayak Production Association, Ozyorsk, Chelyabinsk region, Russia system. Later in 1964–1965 one more non-circulating water reservoir R-11 was constructed downstream (Fig. 1). Since that moment a complex hydrographic system has been used in the upper reaches of the Techa river consisting of four reservoirs (R-3, R-4, R-10, R-11), protecting dams and by-pass canals, which is called the Techa reservoir cascade (TRC).

In the 1950–1960s upland canals were constructed along the left and the right banks of the reservoirs R-10 and R-11 (a left-bank canal – LBC and a right-bank canal – RBC). The regulated discharge from the Kaslinsky-Irtyashsky lake system is performed using the LBC only, while the Mishelyak river discharge is made via the RBC (Fig. 2).

Currently the LBC is an artificial riverbed of the Techa river, while the RBC is an artificial riverbed of the Mishelyak river. The main purpose of the LBC and the RBC is to release excess clean water from Urtyash and Tatysh lakes, as well as interception and drain from the TRC reservoirs of overland and ground water run-offs from the Techa and the Mishelyak catchment area (Fig. 2).

The nearest populated area to the TRC on the Techa river is Muslyumovo settlement (~2,500 residents as of 01.01.1995) located at a distance of about 50 km from the dam D-11.

2 The Parameters of the TRC Radioactive Contamination

The main parameters of radioactive contamination of the TRC reservoirs are given in Table 1. The water in the Techa reservoir cascade is considered to be low-level waste (LLW). At that the major part of the radionuclides is deposited in bottom sediments. The specific activity of the radionuclides in the reservoir R-11 water exceeds the established Russian sanitary regulations only for ^{90}Sr.
LRW releases (~200,000 m^3/year, ~800 Ci/year) are performed only into the reservoirs R-3 and R-4.

The reservoirs R-3, R-4 and R-10 are used as circulating water reservoirs at practically constant water levels.

3 The Components of the TRC Water Balance

During the whole operating period of the TRC depending on the seasonal variations, a stable tendency for water level growth in the reservoir R-11 is observed (Fig. 3).

The level variation in the reservoir R-11 is determined by the inflow and outflow components of the water balance.

The analysis of the available information enabled estimation of the average annual structure of the reservoir R-11 water balance for the period from 1993 till 2006 (Table 2).

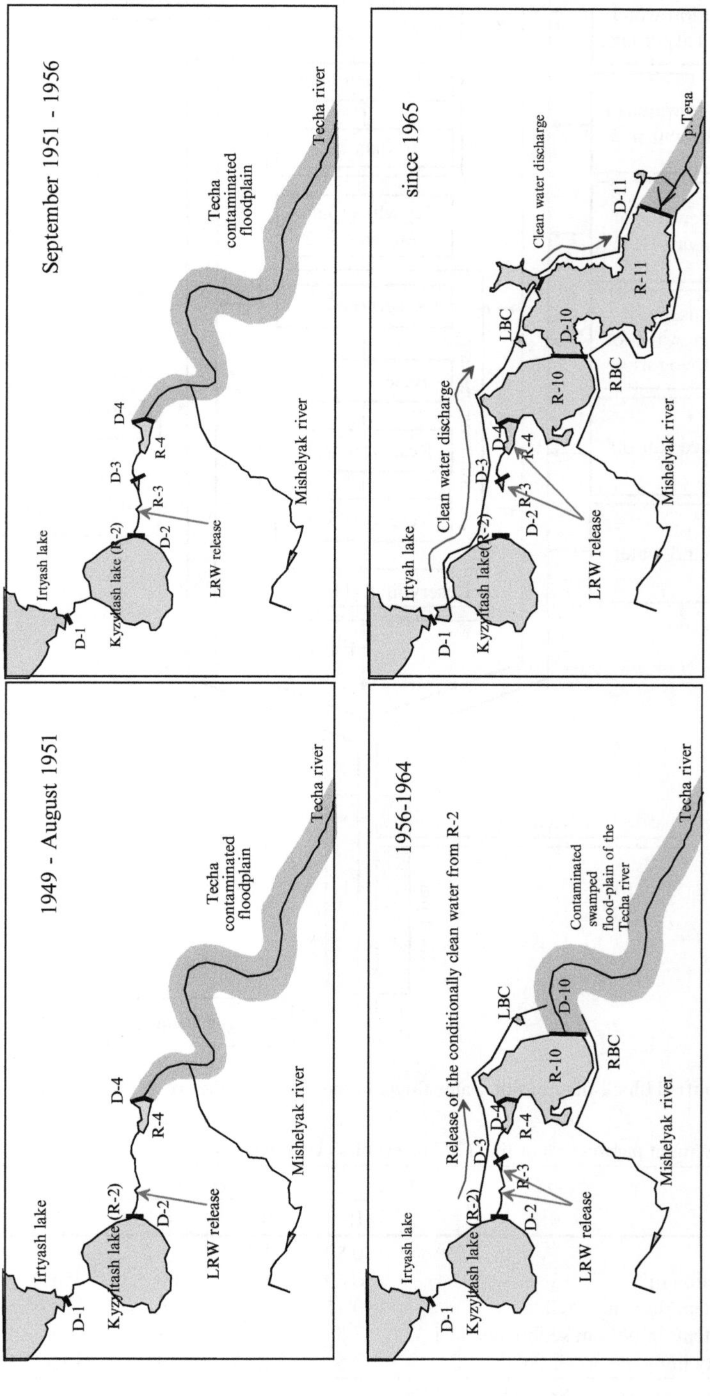

Fig. 1 Stages of construction and reconstruction of the Techa reservoir cascade in the upper Techa river

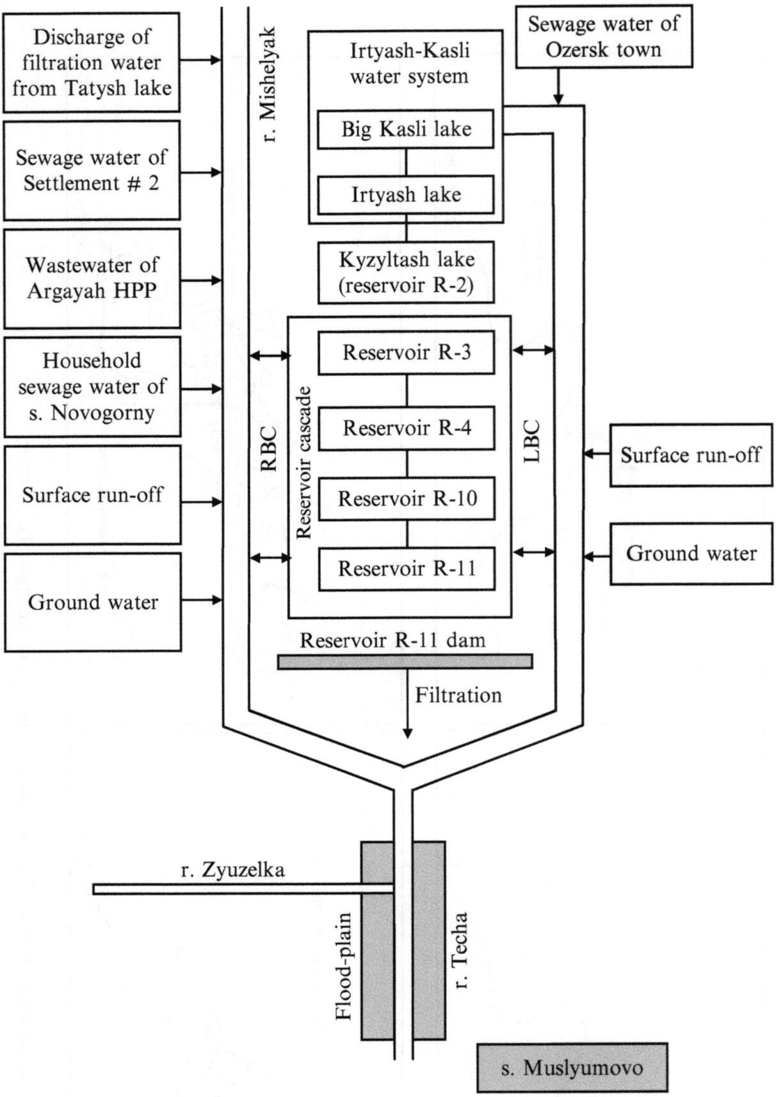

Fig. 2 Simplified block-diagram of water flows in the upper reaches of the Techa river

Table 1 The main parameters of the TRC reservoirs (year 2007)

Parameter	Reservoirs R-3	R-4	R-10	R-11	Total
Area, km^2	0.80	1.3	18.2	46.6	67
Volume, million m^3	0.80	3.8	80	256	341
Activity content[a] in water, kCi	0.15	1.0	16	21	38.1
Activity content[a] in bottom sediments, kCi	7.5	14.0	73	24	118.5
Total[a] activity in the reservoir, kCi	7.6	15.0	89.0	45.0	156.6

[a]The activity is caused by $^{90}Sr + ^{90}Y$ and ^{137}Cs

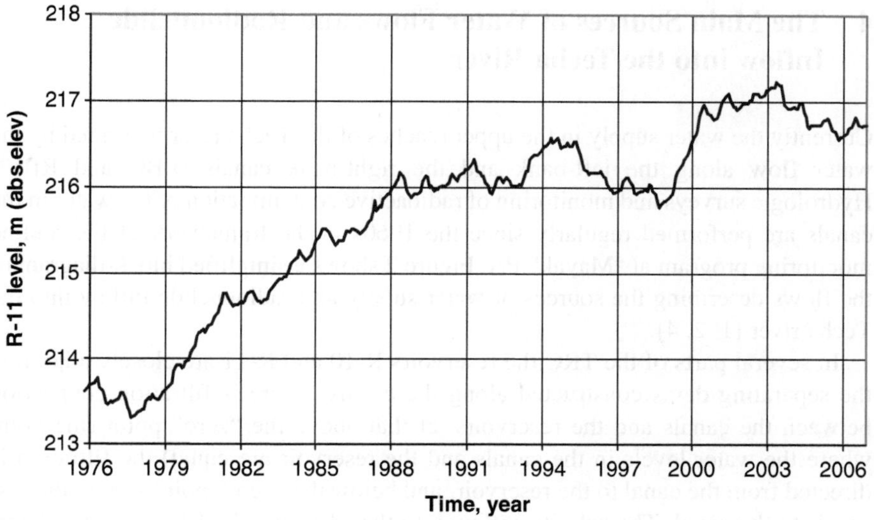

Fig. 3 Water level variation in the R-11 during 1976–2006

Table 2 Average annual structure of the TRC water balance during 1993–2006

Water balance component	Volume (million m^3/year)
1. Inflow components:	
– Low-level waste discharges	0.3
– Sewage and storm water discharge	5.9
– Precipitation	31.4
– Overland and groundwater flow	4.5
Total	42.1
2. Outflow components:	
– Filtration from the TRC	13.4
– Evaporation	23.4
Total	36.8
3. Debalance:	5.3

As the inflow component of the water balance exceeds the outflow one, there is a level growth of the final reservoir – R-11.

The observed tendency for water increase in the reservoir R-11 is caused by the change in meteorological conditions in the vicinity of the enterprise. While prior to 1980 water evaporation from the reservoir surface exceeded precipitation level by ~100 mm/year, after 1980, on the contrary, the precipitation exceeds the evaporation by ~90 mm/year [4].

To stabilize the reservoir R-11 level the inflow component of the water balance should be reduced. The priority solution is to drain waste and sewage water of the enterprise from the TRC and discharge of the clean water into the LBC.

4 The Main Sources of Water Flows and Radionuclide Inflow into the Techa River

Currently the water supply in the upper reaches of the Techa river is formed by the water flow along the left-bank and the right-bank canals (LBC and RBC). Hydrologic surveys and monitoring of radioactive contamination of the water in the canals are performed regularly since the 1960s in the framework of the routine monitoring program at "Mayak" PA. Figure 2 shows a simplified block-diagram of the flows describing the sources of water supply and radionuclide inflow into the Techa river [1, 2, 4].

In several parts of the TRC the reservoirs R-10 and R-11 are closely adjoining the separating dams constructed along the canals. There is filtration connection between the canals and the reservoirs, at that above the "zero" point (the point where the water levels in the canals and the reservoir are equal) the filtration is directed from the canal to the reservoir, and below the "zero" point – from the reservoir to the canal. The activity inflow into the LBC and the RBC water is determined by filtration and sorption properties of the soils and by the difference of water levels in the reservoirs R-10, R-11 and the canals. The main filtration flow runs along the rock under the foundation of the canal dams at a depth of 30–40 m.

Part of the water filtrated through the dam D-11 is intercepted by the drainage system and comes back to the reservoir R-11. The other part of water filtrated under the dam body and through the reservoir bottom is not intercepted by the drainage system and finally is discharged into the Techa river. According to the current estimates total filtration losses from the reservoir R-11 can achieve 10–15 million cubic meters per year. ^{90}Sr specific activity in the reservoir R-11 water is 1,300–1,500 Bq/l, therefore, ~400–600 Ci of ^{90}Sr (15,000–22,000 Bq) annually leaves the reservoir with the filtrate, but inflow into the open hydrographic system of the Techa river is only 23–65 Ci/year (2000–2005) due to ^{90}Sr sorption in the soils. Average specific activity of tritium (^{3}H) in the reservoir R-11 water (2000–2005) is ~900 Bq/l, therefore, the annual inflow into the ground water and directly into the Techa river water is ~240–360 Ci/year (Table 3).

Inflow of ^{137}Cs and plutonium into the Techa river with the water filtrated from the TRC reservoirs is negligible and cannot be detected with the radiation monitoring methods currently used at the "Mayak" PA for the following reasons:

- Specific activity of ^{137}Cs and the sum of plutonium isotopes (ΣPu) in the reservoir R-11 water does not exceed ~3 Bq/l and ~0.005 Bq/l, correspondingly [1].
- High sorption capacities of ^{137}Cs and plutonium limit migration capacity of these radionuclides, therefore, a shielding barrier in the form of loamy lateral protection dam (total width 10–15 m) is practically impenetrable.

The results of numerous experimental studies demonstrate that the most part of activity of long-lived radionuclides (^{90}Sr, ^{137}Cs, ΣPu, ^{99}Tc) is deposited in the

Table 3 Characteristics of the sources of ^{90}Sr and ^{3}H filtration inflow from the TRC into the Techa river water (2000–2005)

Parameter	Sources of ^{90}Sr and ^{3}H filtration inflow			Total
	LBC	RBC	Filtrate under the dam body D-11	
Filtrated water volume, million m^3/year	3–6	2–3	5–8	10–15
Inflow with the filtrate into the Techa river water, Ci/year (thousand Bq/year): ^{90}Sr	10–40 (0.4–1.5)	13–30 (0.5–1.1)	0	23–65 (0.9–2.4)
^{3}H	70–140 (2.7–5.4)	50–70 (1.8–2.7)	120–190 (4.5–7.2)	240–360 (9–14)

swamped river flood-plain (Asanovsky swamps), located from the dam D-11 to Muslyumovo settlement. In the period of LRW releases into the Techa river (1949–1956) the swamped parts of the upper river were strongly retaining the discharged activity serving as a certain natural filter. The activity accumulation in the Asanovsky swamps occurred as a result of sorption processes (for soluble in water radionuclide forms) and deposition of natural and man-made suspensions contaminated with radionuclides on the bottom. In the later period of time (beginning with 1957) the swamped parts of the river presented the main internal source of the secondary contamination of the river water as a result of desorption processes (for ^{90}Sr) and water erosion (for ^{137}Cs, Pu and ^{99}Tc) [3].

Table 4 provides data on inventory of the main long-lived radionuclides deposited in the bottom sediments and the river flood-plain in the upper and lower parts.

The main external factors determining the water specific activity and radioactive run-off of the Techa river (in liquid and solid forms) are water run-off conditions and the source rate of activity inflow with water through hydraulic engineering structures (HES) (LBC and RBC).

Tables 5 and 6 provide characteristics of the sources of water feed and ^{90}Sr inflow into the LBC and the RBC obtained by means of direct instrumental measurements or calculation estimate method for the period 2000–2006.

Analysis of the data provided in Tables 5 and 6 demonstrates that the main sources of the river water infiltration are as follows: water drawdown from Irtyash lake (up to 60–70%), overland and underground inflow (up to 10–20%) and man-made household water discharges (up to 10%). As a result of water filtration from TRC through and under protecting dams into the LBC and RBC there is an inflow of 4–8 million cubic meters per year (~5% from the total water run-off) into the Techa river water, but it is this source that provides major (up to 95%) ^{90}Sr inflow into the river water.

Table 4 Estimate of radionuclide inventory deposited on the different parts of the Techa river system (2005) [3] in kCi (thousand Bq)

	River system part	
Radionuclide	Dam D-11 – s. Muslyumovo	s. Muslyumovo – estuary
^{90}Sr	520 (19)	30 (1.1)
^{137}Cs	4,200 (155)	450 (17)
ΣPu	18.5 (0.68)	3.0 (0.11)[a]
^{99}Tc	8.4 (0.31)	0.72 (0.027)

[a]About 50% of the contained plutonium has global origin

Table 5 Parameters of the main sources of water infiltration and ^{90}Sr inflow into the LBC water for the period from 2000 to 2006 [3]

	Inflow	
Source	water (million m^3/year)	^{90}Sr (Ci/year)
Water pass-by from Irtyash lake	15–160	0.002–0.2
Discharge of household water of Ozersk (Irtyash lake)	14–17	~0.02
Overland and underground run-off from the contaminated area	5–30	0.5–2.0
Filtrate from the TRC reservoirs	3.0–6.0	10–40
Natural weir regulating impact	±0.1	±10
Total	46–242	10–43

Table 6 Parameters of the main sources of water feed and ^{90}Sr inflow into the RBC (the Mishelyak river) during 2000–2006

	Inflow	
Source	water (million m^3/year)	^{90}Sr (Ci/year)
Discharge of household water from settlement #2, Ozersk (B. Akulya lake)	0.4–0.6	~0.003
Filtrate of Tatysh lake (reservoir R-6)	0.3–0.7	~0.06
Discharge of household water of s. Novogorny (underground water intake)	~2.5	<0.015
Filtrate of ash-disposal area of Argayash HPP (Ulagach lake)	~4.5	~0.06
Overland and underground run-off from the contaminated area	2–10	0.1–0.5
Filtrate from the TRC reservoirs	2–3	12–30
Total	14–22	13–32

5 Problems of the TRC Safe Usage

5.1 *Strontium-90 Inflow into the Open Hydrographic System of the Techa River*

The most significant impact of the TRC reservoirs on the environment is made due to radioactive strontium-90 inflow into the open hydrographic system of the Techa river with the filtration flow.

The investigations performed demonstrate (Fig. 4) that dependence of the total strontium-90 filtration inflow from the reservoir R-11 into the LBC and the RBC on the water level in the reservoir has a distinct non-linear character, and it rises from 5–7 Ci/year (at water level in the reservoir 215.5 m) to ~60 Ci/year (at the level 217.0 m).

The radioactive contamination of the water in the Techa river in Muslyumovo settlement area (the nearest to the enterprise populated area on the Techa river) is caused only by the ^{90}Sr content and is determined by the set of the following factors [1, 2]:

- ^{90}Sr inflow into the upper Techa river through hydraulic engineering structures as a result of water filtration from the TRC reservoirs (mainly from the reservoir R-11) into the by-pass canals (see Fig. 4)
- ^{90}Sr sorption/desorption at the swamped part of the river (Asanovsky swamps) located between the HES (reservoir R-11 dam) and Muslyumovo settlement
- Water run-off of the river, which increases significantly in "high-water" years due to additional discharge of "clean" water from Irtyash lake through LBC (see Table 5)

Average annual specific activity of ^{90}Sr in the Techa river water (s. Muslyumovo) makes (1995–2006) 10–15 Bq/l, varying from 2–3 Bq/l during spring flood to 20–40 Bq/l in summer low water and other radionuclide content is always significantly lower than the established Russian sanitary standards.

From 1970 to 1995 strontium run-off in the middle part of the river (s. Muslyumovo) was three to five times higher than in the upper reaches (Fig. 5). In this period of time up to 70–90% ^{90}Sr activity run-off in the middle and the lower river parts was

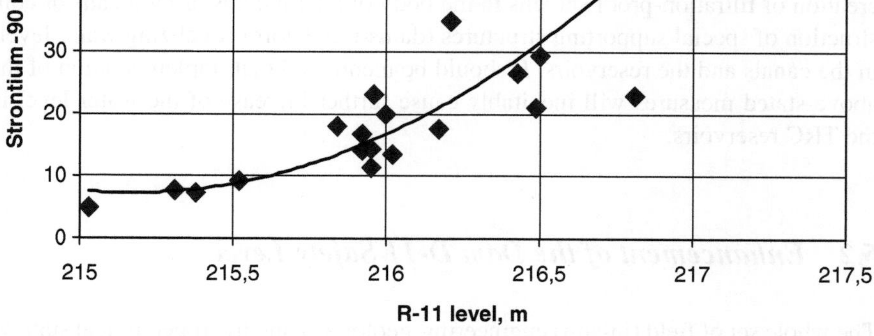

Fig. 4 Dependence of the total filtration inflow of ^{90}Sr in the upper Techa river through LBC and RBC (Ci/year) on the water level in the reservoir R-11

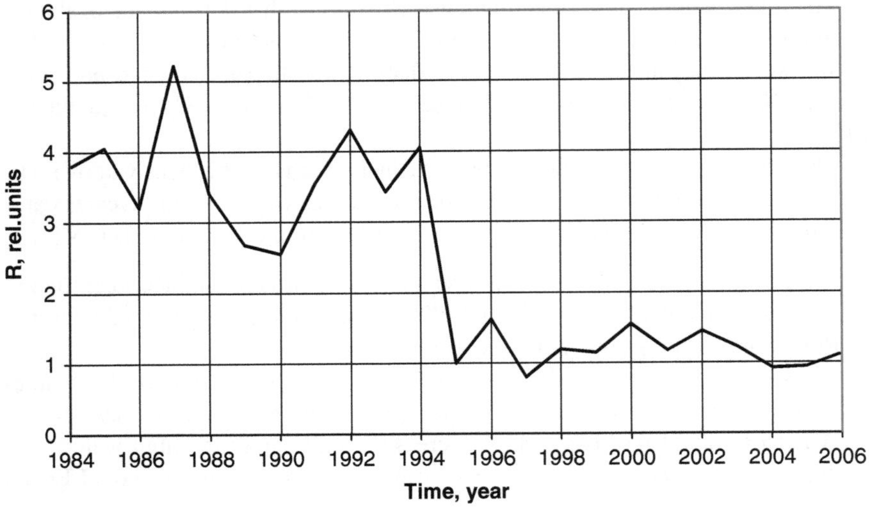

Fig. 5 Ratio of ^{90}Sr run-off R in the hydrosection of s. Muslyumovo to the total ^{90}Sr run-off through LBC and RBC

caused by the activity washout from the Asanovsky swamps, and the total inflow through HES (LBC, RBC and the dam D-11 filtrate) did not exceed 30%. At that ^{90}Sr specific activity in the Techa river water practically did not depend on this radionuclide inflow from the HES, but was determined mainly by the current ^{90}Sr inventory in the swamped part of the river upper reaches (Asanovsky swamps) formed in the 1950s at LRW releases into the Techa river. Beginning with the mid 1990s ^{90}Sr content in the Asanovsky swamps was reduced to such a level (500 Ci at the end of 2005) that radioactive contamination in the river (s. Muslyumovo) started to be determined mainly by the ^{90}Sr filtration inflow through the HES [3].

^{90}Sr inflow into the open hydrographic system of the Techa river open hydrographic system can be significantly reduced, if the volume of filtration drains from the TRD into the LBC and RBC is decreased. This can be achieved by means of creation of filtration-proof screens in the body of lateral dams or by means of construction of special supporting structures (dams-regulators) equalizing water levels in the canals and the reservoirs. It should be mentioned that implementation of the above-stated measures will inevitably cause further increase of the water level in the TRC reservoirs.

5.2 *Enhancement of the Dam D-11 Safety Level*

The whole set of field (in-situ) engineering-geological and hydrogeological surveys was performed in 2000–2003 to obtain the data needed for assessment of the current status and stability of the dam D-11.

Based on the investigation results the status and stability of the reservoir R-11 dam have been assessed. As a result of the investigations performed a general conclusion was made that the dam D-11 met all the requirements for the second reliability class constructions in terms of static, seismic (at earthquake intensity up to seven points) and filtration strength and stability. The dam D-11 stability is ensured for the second-class reliability construction and at the water level increase up to the maximum mark 218.14 m (design reference mark of accidental discharge threshold).

The detailed survey performed in 2003 revealed the abnormal area of water-bearing soils at one of the parts of the dam crest. For elimination of the detected water-bearing areas in the dam D-11 body and prevention of similar phenomena in the future, with the objective to increase the dam safety level, the decision was made to construct a water-proof barrier ("wall in the soil") in the upper part of the dam body.

The activities to construct the filtration-proof screen in the dam D-11 body were started in 2006 and should be finished mainly in 2007 providing the opportunity to increase the reservoir R-11 free capacity by ~40 million cubic meters.

6 The Concept of the TRC Long-Term Usage

For the upcoming 100 years the two main periods of usage, further disposal and rehabilitation of the TRC should be considered:

- During the first (transitional) period lasting for 5–10 years the complete termination of all releases into the reservoirs is planned as well as stabilization of the reservoir levels.
- The second period is connected with ensuring long-term controlled and safe storage (disposal) of the LRW accumulated in the TRC reservoirs, implementation of engineering solutions and detail designs for decreasing of the water levels up to the required ones and radiation rehabilitation of the adjacent territories, as well as minimizing the filtration water inflow into the open hydrographic system of the Techa river.

After the first-priority activities aimed at stabilization of the water level in the TRC are finished, the problem will be studied related to gradual disposal of the reservoir R-3 being the most contaminated TRC reservoir. Elimination of the reservoir water area will be performed using covering technique and construction engineering, tested and used at covering of the reservoir R-9 water area (Karachay). Finally, the reservoir R-3 should be transformed into the solid radioactive waste (SRW) near-surface burial.

The problem of need and methods of the reservoir R-4 elimination (disposal) should be studied and solved after the R-3 water area covering. The reservoirs R-10 and R-11 will be used under disposal conditions (long-term storage of the accumulated activity) during 50–100 years at the minimum.

List of Abbreviations

LRW liquid radioactive waste
TRC Techa reservoir cascade
LBC left-bank canal
RBC right-bank canal
LLW low-level waste
(ΣPu) sum of plutonium isotopes
HES hydraulic engineering structures
SRW – solid radioactive waste

References

1. Joint Norwegian-Russian Expert Group, Malyshev S V, Vakulovsky S M, Drozhko E G, Romanov G N, Glagolenko Yu V, Mokrov Yu G, Westerlund E A, Amundsen I, Strand P, Salbu B, Oughton D H, Christensen G C, Bergan T S (1997) Sources contributing to radioactive contamination of the Techa River and areas surrounding the "MAYAK" production association, Urals, Russia. Osteras: 134.
2. Joint Norwegian-Russian Expert Group, Malyshev S V, Glagolenko Yu V, Drozhko E G, Mokrov Yu G, Romanov G N, Stukalov P M, Vakulovsky S M, Novitsky M A, Amundsen I, Strand P, Bergan T S, Kozlov R, Iosjpe M, Standring W J F, Bborghuis S, Salbu B, Oughton D H, Skipperud L, Mobbs H, Tronstad E, Christensen G C, Varskog P (2004) Impacts on man and the environment in northern areas from hypothetical accidents at "Mayak" PA, Urals, Russia. Program on Investigations of Possible Impacts of the "Mayak" PA Activities on Radioactive Contamination of the Barents and Kara Seas. Osteras: 104.
3. Glagolenko Yu V, Drozhko E G, Mokrov Yu G (2007) Specific features of formation of radioactive contamination of the Techa river. Radiat Saf Probl 2: 27–36.
4. Sadovnikov V I, Glagolenko Yu V, Drozhko E G, Mokrov Yu G, Stukalov P M (2002) Current status and solutions of the Techa reservoir cascade problems. Radiat Saf Probl 1: 3–14.

Restoration Principles and Criteria: Superfund Program Policy for Cleanup at Radiation Contaminated Sites

S. Walker

Abstract The United State's Environmental Protection Agency (EPA) Office of Superfund Remediation and Technology Innovation (OSRTI) is responsible for implementing the long-term (non-emergency) portion of a key U.S. law regulating cleanup: the Comprehensive Environmental Response, Compensation and Liability Act, CERCLA, nicknamed "Superfund." The purpose of the Superfund program is to protect human health and the environment over the long term from releases or potential releases of hazardous substances from abandoned or uncontrolled hazardous waste sites. The focus of this paper is on Superfund, including how radiation is addressed by the Superfund program.

This paper provides a brief overview of the approach used by EPA to conduct Superfund cleanups at contaminated sites, including those that are contaminated with radionuclides, to ensure protection of human health and the environment. The paper addresses how EPA Superfund determines if a site poses a risk to human health and the framework used to determine cleanup levels. The theme emphasized throughout the paper is that within the Superfund remediation framework, radioactive contamination is dealt with in a consistent manner as with chemical contamination, except to account for the technical differences between radionuclides and chemicals. This consistency is important since at every radioactively contaminated site being addressed under Superfund's primary program for long-term cleanup, the National Priorities List (NPL), chemical contamination is also present.

Keywords Superfund Baseline Risk Assessment, EPA, remediation framework, radioactive contamination, chemical contamination, remedy selection, Health Effects Assessment Summary Tables (HEAST), PRG

Science and Policy Branch, Office of Superfund Remediation and Technology Innovation (OSRTI), Environmental Protection Agency, USA

1 Introduction: Superfund Based on Protection of Human Health

The United State's Environmental Protection Agency (EPA) Office of Superfund Remediation and Technology Innovation (OSRTI) is responsible for implementing the long-term (non-emergency) portion of a key U.S. law regulating cleanup: the Comprehensive Environmental Response, Compensation and Liability Act, CERCLA, commonly known as "Superfund." The purpose of the Superfund program is to protect human health and the environment over the long term from releases or potential releases of hazardous substances from abandoned or uncontrolled hazardous waste sites. The focus of this paper is on Superfund, including how radiation is addressed by the Superfund program.

This paper provides a brief overview of the approach used by EPA to conduct Superfund cleanups at contaminated sites, including those that are contaminated with radionuclides, to ensure protection of human health and the environment. The paper addresses how EPA Superfund determines if a site poses a risk to human health and the framework used to determine cleanup levels. The theme emphasized throughout the paper is that within the Superfund remediation framework, radioactive contamination is dealt with in a consistent manner as with chemical contamination, except to account for the technical differences between radionuclides and chemicals. This consistency is important since at every radioactively contaminated site being addressed under Superfund's primary program for long-term cleanup, the National Priorities List (NPL), chemical contamination is also present.

The Superfund program is dedicated to cleaning up hazardous waste sites and protecting public health and the environment. EPA has worked closely with the Agency for Toxic Substances and Disease Registry (ATSDR) in evaluating the impacts of Superfund sites on public health. Studies conducted by the ATSDR show a variety of health effects associated with some Superfund sites, including birth defects, cardiac disorders, changes in pulmonary function, impacts on the immune system, infertility, and increases in chronic lymphocytic leukemia. In addition, hundreds of drinking water wells across the country have been shut down due to contamination.

Superfund is one of the United States' most ambitious and complex environmental programs. Since it was launched, the Superfund program has maintained two bedrock principles: protection of human health and the environment is foremost, and polluters must pay for cleanup of the contamination they create. Superfund arose out of the need to protect citizens from the dangers posed by abandoned or uncontrolled hazardous waste sites. In the wake of the discovery that a residential district had been built atop an abandoned chemical dump at a town called "Love Canal" in New York State, the American public demanded that its government take action.

The enactment of Superfund gave the federal government broad authority to respond to hazardous substance emergencies, and to develop long-term solutions for the United States' most serious hazardous waste problems like Love Canal. It also enabled the United States Federal government to recover the costs from responsible parties, or to force them to clean up the hazardous site at their own expense. When CERCLA or Superfund was enacted, the challenge of what was assumed to be a few hundred discrete, land-based cleanups appeared relatively straightforward. Furthermore, the Congress created a $1.6 billion Trust Fund to ensure that funding would prove no obstacle. Things have not worked out as smoothly as that, however. The problem of neglected hazardous waste sites has revealed itself to be far more complex and widespread than anyone at first realized.

While every Superfund site is unique, and thus cleanups must be tailored to the specific needs of each site, there are two requirements that must be met at every site. CERCLA requires that all remedial actions at Superfund sites must be protective of human health and the environment. Therefore, cleanup actions are developed with a strong preference for remedies that are highly reliable, provide long-term protection and provide treatment of the principle threat to permanently and significantly reduce the volume, toxicity, or mobility of the contamination. In addition, EPA believes that site cleanups should protect ground waters that are current or potential sources of drinking water to drinking water standards whenever practicable. In addition, CERCLA specifically requires Superfund actions to attain or waive the standards and requirements found in other State and Federal environmental laws and regulations. This mandate is known as compliance with "applicable or relevant and appropriate requirements" or ARARs.

2 Remedy Selection

A comprehensive regulation known as the National Oil and Hazardous Substances Pollution Contingency Plan or NCP contains the guidelines and procedures for implementing the Superfund program. The NCP reiterates CERCLA's goal of selecting remedies that protect human health and the environment, that maintain protection over time, and that minimize untreated waste. The NCP sets forth nine criteria for selecting Superfund remedial actions. These evaluation criteria are the standards by which all remedial alternatives are assessed and are the basis of the remedy selection process. The criteria can be separated into three levels: threshold, balancing, and modifying. The first two criteria are known as "threshold" criteria. They are a reiteration of the CERCLA mandate that remedies must (1) at a minimum assure protection of human health and the environment and (2) comply with (or waive) requirements of other Federal environmental laws, more stringent State environmental laws and State facility-siting laws. They are the minimum requirements that each alternative must meet in order to be eligible for selection as a remedy.

After the threshold criteria are applied, EPA considers a number of other evaluation criteria. Five of the criteria are known as the "balancing" criteria. These criteria are factors with which tradeoffs between alternatives are assessed so that the best option will be chosen, given site-specific data and conditions. The criteria balance long-term effectiveness and permanence; reduction of toxicity, mobility, or volume; short-term effectiveness; implementability; and cost. The final two criteria are called "modifying" criteria: new information or comments from the State or the community may modify the preferred remedial action alternative or cause another alternative to be considered.

EPA believes the "modifying" criteria concerning new information or comments from the local community is important. In many instances, communities are able to provide valuable information on local history, citizen involvement, and site conditions. To ensure community participation, EPA specifically requires the party conducting the cleanup to conduct a number of activities. For example, EPA conducts community interviews and develops a community relations plan to help EPA determine the community's level of interest in the site, major concerns and issues. EPA creates an information repository and administrative record for every site and makes it available to community members. EPA also develops a document specifically for the community which explains the various clean up options being considered, holds at least one meeting to explain the options and invites the community to submit comments on them. EPA also make funding available to eligible community members so they may obtain technical assistance to better understand the often complex issues associated with cleaning up a Superfund site. By identifying the public's concerns, EPA is able to fashion a response that more effectively addresses the community's need.

3 Risk Assessment

To help meet the Superfund program's mandate to protect human health and the environment from current and potential threats posed by uncontrolled hazardous substance releases (both radiological and nonradiological), EPA has developed a human health evaluation process as part of its remedial response program. The process of gathering and assessing human health risk information is adapted from well-established chemical risk assessment principles and procedures. The Superfund Baseline Risk Assessment provides the EPA's estimate of the likelihood and magnitude of health problems occurring if no cleanup action is taken at a site. Specifically, the risk assessment provides:

- An analysis of baseline risks to help determine the need for action at sites
- A basis for determining levels of hazardous substances that can remain onsite and still be adequately protective of public health
- A basis for comparing potential health impacts of various remedial alternatives and
- A consistent process for evaluating and documenting public health threats at sites nationwide

The results of a risk assessment are critical in determining whether responses to protect human health and the environment are justified, and in establishing an appropriate cleanup level. The risk assessment also helps EPA identify potential risks associated with a particular remedy and evaluate risks remaining at a site after cleanup is completed.

4 Risk-Based Cleanup Levels

Cleanup levels for radioactive contamination at CERCLA sites are generally expressed in terms of risk levels, rather than millirem or millisierverts, as a unit of measure. CERCLA guidance recommends the use of slope factors in the EPA Health Effects Assessment Summary Tables (HEAST) when estimating cancer risk from radioactive contaminants, rather than converting from millirem. HEAST is based on risk coefficients in Federal Guidance Report 13.

Compliance with the requirements of other Federal environmental laws, more stringent State environmental laws, or State facility-siting laws is often the determining factor in establishing cleanup levels at CERCLA sites. These requirements are known as Applicable or Relevant and Appropriate Requirements (ARARs). However, where ARARs are not available or are not sufficiently protective, EPA generally sets site-specific remediation levels for: (1) carcinogens at a level that represents an upper-bound lifetime cancer risk to an individual of between 10^{-4} to 10^{-6}; and for (2) non-carcinogens such that the cumulative risks from exposure will not result in adverse effects to human populations (including sensitive sub-populations) that may be exposed during a lifetime or part of a lifetime, incorporating an adequate margin of safety. The specified cleanup levels account for exposures from all potential pathways, and through all media (e.g., soil, ground water, surface water, sediment, air, structures, biota).

The 10^{-4} to 10^{-6} cancer risk range can be interpreted to mean that a highly exposed individual may have a 1 in 10,000 to 1 in 1,000,000 increased chance of developing cancer because of exposure to a site-related carcinogen. Once a decision has been made to take an action, EPA prefers cleanups achieving the more protective end of the range (i.e., 10^{-6}). EPA uses 10^{-6} as a point of departure and establishes Preliminary Remediation Goals (PRGs) at 1×10^{-6}.

To assess the potential for cumulative noncarcinogenic effects posed by multiple contaminants, EPA has developed a hazard index (HI). The HI is derived by adding the noncancer risks for site contaminants with the same target organ or mechanism of toxicity. When the HI exceeds 1.0, there may be concern for adverse health effects due to exposure to multiple contaminants. Radioisotopes of uranium are generally the only radionuclides for which EPA will evaluate the HI.

4.1 *PRGs*

PRGs are used for site "screening" and as initial cleanup goals if applicable. PRGs are not de facto cleanup standards and should not be applied as such. The PRG's role in site "screening" is to help identify areas, contaminants, and conditions that do not require further federal attention at a particular site.

PRGs not based on ARARs are risk-based concentrations, derived from standardized equations combining exposure information assumptions with EPA toxicity data. PRGs based on cancer risk are established at 1×10^{-6}. PRGs are identified early in the CERCLA process. PRGs are modified as needed based on site-specific information.

4.2 Superfund Risk and Dose Soil and Water Models

EPA has developed a PRG for Radionuclides electronic calculator, known as the Rad PRG calculator. This electronic calculator presents risk-based standardized exposure parameters and equations that should be used for calculating radionuclide PRGs for residential, commercial/industrial, and agricultural land use exposures, tap water and fish ingestion exposures. The calculator also presents PRGs to protect groundwater which are determined by calculating the concentration of radioactively contaminated soil leaching from soil to groundwater that will meet MCLs or risk-based concentrations. The Rad PRG calculator may be found at: http://epa-prgs.ornl.gov/radionuclides/.

To address ARARs that are expressed in terms of millirem per year, an approach similar to that taken for calculation of PRGs was also used to calculate soil "compliance concentrations" based upon various methods of dose calculation in another EPA tool, the "Dose Compliance Concentrations", or DCC calculator The DCC calculator equations are identical to those in the PRG for Radionuclides, except that the target dose rate (ARAR based) is substituted for the target cancer risk (1×10^{-6}), the period of exposure is one year to indicate year of peak dose, and a DCF will be used in place of the slope factor. The DCC calculator may be found at: http://epa-dccs.ornl.gov/.

4.3 Superfund Decommissioning Models

EPA has recently completed one risk assessment tool, and is close to completion of another that are particularly relevant to decommissioning activities conducted under CERCLA authority. EPA developed the Preliminary Remediation Goals for Radionuclides in Buildings (BPRG) electronic calculator to help standardize the evaluation and cleanup of radiologically contaminated buildings at which risk is being assessed for occupancy. BPRGs are radionuclide concentrations in dust, air and building materials that correspond to a specified level of human cancer risk. The BPRG calculator may be found at: http://epa-bprg.ornl.gov/.

The intent of the draft Preliminary Remediation Goals for Radionuclides in Outside Surface SPRG calculator is to address hard outside surfaces such as building slabs, outside building walls, sidewalks and roads. SPRGs are radionuclide concentrations in dust and hard outside surface materials.

4.4 Superfund Ecological Risk Model

EPA is also developing the "Radionuclide Ecological Benchmark" calculator. This calculator provides biota concentration guides (BCGs), also known as ecological screening benchmarks, for use in ecological risk assessments at CERCLA sites. This calculator is intended to develop ecological benchmarks as part of the EPA guidance "Ecological Risk Assessment Guidance for Superfund: Process for Designing and Conducting Ecological Risk Assessments." The calculator develops ecological benchmarks for ionizing radiation based on cell death only.

5 Compliance with Environmental Laws

Compliance with (or waiver of) requirements of other Federal environmental laws, more stringent State environmental laws and State facility-siting laws is a cornerstone of CERCLA. Cleanups conducted under the Superfund program must comply with these laws unless a waiver is justified. These laws, as well as ARARs, assist EPA in identifying preliminary remediation goals and alternatives. Complying with ARARs both during the implementation and upon completion of an action helps the lead agency define the ways in which the activity can be carried out in a manner that is protective of human health and the environment.

Because the diverse characteristics of Superfund sites preclude the development of prescribed ARARs, it is necessary to identify ARARs on a site-by-site basis. There are many radiation standards that are likely to be used as ARARs to establish cleanup levels or to conduct remedial actions. Some of the radiation standards most frequently used as ARARs at Superfund sites are the soil cleanup and indoor radon standards developed to address contamination at sites that are subject to the Uranium Mill Tailings Radiation Control Act of 1978 (UMTRCA). When used as an ARAR at Superfund sites, the soil cleanup level for radium 226 and radium 228 combined, or thorium 230 and thorium 232 combined, is 5 picoCuries per gram (pCi/g) above background, while the indoor radon level is 0.02 working levels inclusive of background. For a list of "Likely Federal Radiation Applicable or Relevant and Appropriate (ARARs)", see Attachment A of EPA's guidance "Establishment of Cleanup Levels for CERCLA sites with Radioactive Contamination" at http://www.epa.gov/superfund/health/contaminants/radiation/pdfs/radguide.pdf.

6 Groundwater

One extremely important ARAR that should be noted are Maximum Contaminant Levels (MCLs) that are established under the United States law for drinking water standards, called the Safe Drinking Water Act. Over 85% of the sites designated for long-term cleanup by the Superfund program have some groundwater contamination. Ground water contamination is generally more difficult to detect and clean up than contamination in other environmental media. Ground water generally moves slowly; velocities are usually in the range of 5–50 feet per year. Large quantities of a particular contaminant can enter an aquifer and remain undetected until reaching a point of use, such as a water well or surface water body. Moreover, contaminants in ground water, unlike those in other environmental media such as air or surface water, generally move with relatively little mixing or dispersion, so concentrations can remain relatively high. These plumes of concentrated contaminants move slowly through aquifers and may be present for many years, sometimes for decades or longer, making the resource potentially unusable for extended periods of time. Slow migration over an extended period can cause a large area to become contaminated, and will increase the potential for exposure to those contaminants. All of these factors favor prevention of contaminated ground water where possible, and remediation of chemical and radioactive materials in other media (e.g., soil) to prevent future contamination of ground water.

EPA believes contaminated ground water should be restored to beneficial use, whenever practicable. This means that sites where the contaminated ground water is a potential or current source of drinking water should be remediated to concentrations corresponding to drinking water standards (e.g., concentrations corresponding to MCLs or more stringent State drinking water standards). The Superfund program requires MCLs be met within the aquifer, not at the tap. EPA's phased approach to addressing contaminated groundwater at CERCLA sites is discussed in "Presumptive Response Strategy and Ex-Situ Treatment Technologies for Contaminated Ground Water at CERCLA Sites, Final Guidance, which may be found at: http://www.epa.gov/superfund/health/conmedia/gwdocs/gwguide/index.htm.

EPA's policy is to defer to State determinations of ground-water use when such determinations are based on a Comprehensive State Ground Water Protection Program (CSGWPP) that has (1) been endorsed by EPA and (2) allows such determinations to be made at specific sites. In the absence of a CSGWPP, EPA considers other state classification schemes and EPA's classification guidelines which use criteria defining ground waters of sufficient quantity and quality to supply the needs of a single family household. EPA's us of CSGWPP's at CERCLA sites is discussed in "The Role of CSGWPPs in EPA Remediation Programs" which may be found at: http://www.epa.gov/superfund/health/conmedia/gwdocs/pdfs/role.pdf.

The current MCLs for radionuclides are set at 4 mrem/year to the whole body or an organ for the sum of the doses from beta particles and photon emitters, 15 picoCuries per liter (pCi/l) for gross alpha, and 5 pCi/l combined for radium-228 and radium-226, and 30 µg/l of uranium. EPA has published concentration tables for each radionuclide that correspond to the 4 mrem/year MCL which may be found at:

http://www.epa.gov/safewater/radionuclides/pdfs/guide_radionuclides_table-betaphotonemitters.pdf.

7 Other Criteria, Advisories and Guidance

Many Federal and State environmental and public health agencies develop criteria, advisories, guidance, and proposed standards that are not legally enforceable but contain information that would be helpful in carrying out selected remedies, or in determining their protectiveness. These materials are meant to complement the use of ARARs, not to compete with or replace them. Because they are not ARARs, their identification and use are not mandatory. These are known as to-be-considered (TBC) material. However, it is EPA's policy that dose-based (millirem or millisievert) recommendations should generally not be used as TBCs.

In conjunction with the completion of the baseline risk assessment, where no ARARs address a particular situation, or the existing ARARs do not ensure sufficient protectiveness, these advisories, criteria or guidelines are used to set cleanup targets. This information may be invaluable in deciding how to carry out a particular remedy. Many ARARs have broad performance criteria but do not provide specific instructions for implementation. Often those instructions are contained in supplemental program guidance. Sometimes the Superfund program develops guidance on interpreting a particular ARAR to assist site decision makers. These guidance documents on compliance with ARARs at radioactively contaminated CERCLA sites may be found at the following webpage: http://www.epa.gov/superfund/health/contaminants/radiation/radarars.htm.

8 Land Use/Institutional Controls

The concentration levels for various media that correspond to the acceptable risk level established for cleanup will depend in part on land use at the site. Land uses that will be available following completion of a response action are determined as part of the remedy selection process considering the reasonably anticipated land use or uses along with other remedy selection factors. EPA's policies for how to determine a sites reasonably anticipated land use is discussed in "Land Use in the CERCLA Remedy Selection Process", which may be found at: http://www.epa.gov/superfund/community/relocation/landuse.pdf.

Institutional controls are generally included as a supplemental component to cleanup alternatives, not as a substitute for treatment or containment. Institutional controls are non-engineering measures – usually, but not always legal controls – intended to affect human activities in a way that prevents or reduces exposure to hazardous substances. Institutional controls usually restrict land use to prevent unanticipated changes in use that could result in unacceptable exposures to residual contamination. At a minimum, institutional controls are intended to alert future users to the residual risks and the need to monitor for any changes in use. EPA's

CERCLA policy states that if a site cannot be cleaned up to a protective level (i.e., generally within the 10^{-4} to 10^{-6} risk range) for the "reasonably anticipated future land use" because it is not cost-effective or practicable, then a more restricted land use should be chosen that will meet a protective level.

Where waste is left on-site at levels that would require limited use and restricted exposure to ensure protectiveness, EPA will conduct reviews at least once every five years to monitor the site for any changes including changes in land use. Such reviews need to analyze the implementation and effectiveness of any institutional controls with the same degree of care as other parts of the remedy. Should land use change in spite of land use restrictions, it will be necessary to evaluate the implications of that change for the selected remedy, and whether the remedy remains protective.

9 Closing

Actions under Superfund must result in the protective cleanup of sites. The CERCLA framework for addressing hazardous sites ensures that risks from radiological contamination will be addressed in a manner consistent with risks from non-radiological contamination, except to account for technical differences posed by radionuclides, and that cleanups for all contaminants will achieve protection of human health and the environment. The same set of principles and decision making criteria apply equally to both chemical and radioactive hazards. The goal is to provide lasting, protective site restoration while taking into account the cost and achievability of different approaches to attaining these protective goals.

For more information and copies of EPA guidance documents for addressing radioactively contaminated CERCLA sites, see the EPA's Superfund Radiation webpage at:

http://www.epa.gov/superfund/health/contaminants/radiation/index.htm.

For more information and copies of EPA guidance documents for developing cleanup levels for long-term CERCLA sites, see EPA's Remedy Decisions webpage at

http://www.epa.gov/superfund/policy/remedy/sfremedy/index.htm.

Both of these webpages contain numerous OSWER Directives, which are EPA's official guidance for the Superfund program, and other material that is useful for cleaning up CERCLA sites.

Session IV
Safety Regulation Experience in Russia and Abroad

Chairmen: S. Testov, MoD RF and J. Valentin, ICRP

Challenges in Radiation Safety Regulation with Respect to Supervision of FSUE "SevRAO" Facilities

V.R. Alekseeva

Abstract This paper sets out the link between on-going and planned operations of SevRAO sites, the radiation safety issues related to those sites and the regulations which address these issues.

Keywords FSUE "SevRAO", health protection zone, NRB-99

FSUE «Northern Federal enterprise for radioactive waste management» (hereinafter referred to as FSUE "SevRAO") operates in the field of nuclear energy use. It was arranged in 2000 to perform management of spent nuclear fuel (SNF), solid and liquid radioactive wastes (SRW and LRW, respectively), both accumulated in the course of Navy active operation and generated during dismantlement of nuclear submarines (NS) and nuclear powered above water ships; it also carries out environmental remediation of radiation-hazardous sites.

FSUE "SevRAO" consists of three branches, situated in the coastal strip of Barents Sea: branch No. 1 – Zaozersk city (Andreeva bay), branch No. 2 – Osrtovnoy city (Gremikha), branch No. 3 – Saida bay (first stage). These branches perform:

- No. 1, 2 – management of irradiated fuel assemblies (IFA), as well as of SRW and LRW generated in the course of operation and dismantlement of NS and above water ships equipped with nuclear powered installations (NPI)
- No. 3 – long-term storage of reactor compartments, generated in the course of NS dismantlement

Radiation-hazardous and nuclear-hazardous facilities are:

- In branches No. 1 and No. 2 – storage facilities of SNF and radioactive waste (RW)
- In branch No. 3 – reactor compartments

Regional Department No. 120 of Federal Medical-Biological Agency (FMBA), Snezhnogorsk, Russia

The radiation hazard category of the facilities, according to the provisions of paragraph 3.1 of SP 2.6.1.799-99 «Main sanitary rules of radiation safety assurance (OSPORB-99) » – category 1 for branches No. 1 and 2; for branch No. 3 – category 2.

Main conditions, which specify environmental risks during SevRAO facilities operation, include:

- Availability and conditions of SNF
- Amount and conditions of SRW and LRW
- Technical conditions of the sites
- Radiation situation at the facility, within the health protection zone (HPZ) and in the supervised area (SA)
- Development of infrastructure suitable for performance of operations at the facilities (towboats, dock, structures for SNF and RW management)
- Development of structures for environmental monitoring support, including radiation monitoring
- Supervision and control of the operation performed

FSUE «SevRAO» was taken on sanitary and epidemiological service in January of 2002, following the Order of the Federal Department "Medbioextrem". In 2001, "Medbioextrem's" commissions assessed the sanitary and technical conditions of buildings, constructions and systems of FSUE "SevRAO" branches No. 1 and No. 2 as unsatisfactory.

Large amount of SNF as well as SRW and LRW are stored in Andreeva bay and Ostrovnoy. The very large SNF store facility is situated in Andreeva bay, where in 1960s a base was built for discharge and storage of SNF from Russian Navy nuclear powered ships and submarines. After termination of active operation of the shore technical base (STB), in 1980s, a level of preventive maintenance decreased significantly, therefore, complicated infrastructure was broken considerably. Life time of majority buildings terminated, and some of them are in poor conditions that caused radioactive contamination of some parts of the site.

STBs in Andreeva bay and Gremikha village accepted from RF Ministry of Defense did not meet requirements for nuclear, radiation and environmental safety, were partly or fully broken, so, SNF and RW management became impossible.

There are contaminated parts on the site of temporary storage areas, where dose rate reaches 140–620 μSv/h. On the industrial sites, at some places, man-made contamination of the topsoil is registered with Cs-137 and Sr-90, which is 100 and more times higher than background value typical for this region, and soil contamination levels with Cs-137 are 4–20 times higher than those for Sr-90. According to data of radiation environmental monitoring nearby the site of temporary storage in Andreeva bay, man-made radionuclide contents in seawater are exceeded considerably (in coastal strip of HPZ, in bottom sediments, seaweeds, vegetations and mushrooms growing within HPZ).

Now, due to the foreign investments, large volume of actions had been completed, directed to improvement of conditions of work and radiation safety in branch No. 1:

- The road is restored.
- A system of physical protection is under construction.

- Buildings of control entrance posts, administrative and domestic combine, as well as an office cloakroom for the personnel had been constructed.
- Two mobile and two module decontamination facilities (sanitary passes) had been commissioned.
- Two facilities had been constructed and equipment had been purchased for decontamination operations and LRW management.
- Equipment for radiation monitoring (RM) had been purchased, including personal dosimeters, spectrometers, portable and stationary RM instruments.
- Three hundred facility specific SRW containers had been manufactured.
- Industrial-laboratory building had been commissioned after completion of overhaul; which is ready for allocation of equipment and performance of relevant radiochemical, radiometric, and spectrometric investigations within class 2 operations; laundry for the special clothes without radioactive contamination; offices.
- The first stage of ASCRO system had been commissioned, which has gamma background detectors and devices for air monitoring of aerosols.
- Water line had been constructed.

However, the process of SNF and RW management, as well as remediation of sites will be long-term and require not only building of the relevant infrastructure, but also arrangement of effective supervision of compliance with radiation safety requirements. With the purpose of radiation safety of workers, the public and environment, special attention in the course of supervision is paid to taking a set of sanitary – hygienic, sanitary – technical and specific medical actions, including:

1. Specification of radiation hazard category of the site, as well as definition of HPZ and SA borders
2. Availability of authorization for operations using radiation sources (sanitary and epidemiological certificates, licenses)
3. Subdivision the industrial site area into some radiation-hygienic zones, according to the level of hazard
4. Elaboration of measures for prevention of unauthorized access to radiation sources (RS), organization of sanitary warrant regime and control of sanitization of persons working in the controlled area
5. Decontamination of surfaces of equipment, rooms and vehicles
6. System of reference levels of radiation situation parameters
7. Observance of radiation safety instructions at the facility, elaboration of measures assuring radiation safety during all types of operations, including radiation-hazardous
8. Elaboration of measures for prevention of radiological accidents, protection of workers, the public and environment in case of accident
9. Subdivision of workers into personnel A and B groups
10. Education and training of the personnel admitted to operations using RS
11. Availability of regulatory documents with respect to sanitary legislation, including radiation safety issues

12. Arrangement of safe conditions of work with RS, reduction of ionizing radiation levels using protective materials and constructions, operation of the special ventilation system and air purification against radioactive substances
13. A system of collection, interim storage and removal of radioactive substances
14. A set of radiation situation controls in workplaces, in workshops, on-site, within HPZ and SA, radioactive effluents and discharges monitoring
15. Monitoring and account of annual individual doses of the personnel A and B group, as well as cases of excess authorized dose limits
16. Arrangement of the system of personnel and public information provision relating to radiation situation
17. Quality of performance of periodic medical examinations of workers
18. Annual presentation of radiation-hygienic certificate of the facility, record forms of the single state system of control and account of individual doses (ESKID)
19. Arrangement of physical protection system at the facility
20. Composition and organizational structure of the radiation safety service of facilities

The Regional Department No. 120 assures radiation safety at FSUE "SevRAO" facilities in compliance with the RF Constitution, main federal laws (FL) of the Russian Federation (FL «About radiation safety of the public», FL «About nuclear energy use», FL «About sanitary and epidemiological prosperity of the public», FL «About preservation of the environment», Water codex, Forestry codex).

Regulatory and legal acts of the RF Government (RF Governmental Directives «About performance of the state sanitary epidemiological supervision in the Russian Federation», «About approval of the Provisions relating to performance of social hygienic monitoring», «About the procedure of elaboration of radiation hygienic certificates of facilities and sites», «About the procedure of arrangement of the single state system of control and account of individual doses of exposure to citizens», «About the Federal target program "Nuclear and radiation safety of Russia"» etc.).

Federal sanitary rules and norms: SP 2.6.1.758–99 «Norms of radiation safety» NRB-99, SP 2.6.1.799–99 «Main sanitary rules of radiation safety assurance» OSPORB-99, SP 2.6.6.1168–02 Sanitary rules of radioactive waste management SPORO-2002, SanPiN 2.3.2.1078–01 «Hygienic requirements for safety and food cost of foodstuffs» with respect to contents of Cs-137 and Sr-90 in foodstuffs and drinking water», SP HPZ and SA-07 «Health protection zones and supervised areas of radiation facilities. Operational conditions and justification of borders».

Special regulations in the field of radiation safety of nuclear ship-building, dismantlement of ships equipped with NPI, nuclear submarines, which had been developed in SRI of the industrial hygiene and marine medicine of FMBA of Russia during the period 2003–2007:

- SP 2.6.1.2040–05 "Radiation safety assurance during design, construction, operation and decommissioning of nuclear powered ships" (SP RB AS-2005)
- SP 2.6.1.2154–06 "Radiation safety assurance during overall dismantlement of nuclear submarines"

- SP 2.6.5.12–02 "Radiation hygienic requirements for dismantled reactor compartment of nuclear submarines during their preparation for store on land"
- R 2.6.1.35–02 "Radiation safety assurance during discharge of irradiated fuel assemblies of nuclear submarines under dismantlement" (RBV-2002)
- R 2.6.6.42–02 "Radiation hygienic requirements for allocation of solid radioactive waste in reactor compartments of nuclear submarines under dismantlement "
- R 2.6.6.37–02 "Hygienic regulations established in the course of nuclear submarine dismantlement"
- R 2.6.1.25–07 "Criteria and norms for remediation of facilities and sites, contaminated with man-made radionuclides, pertaining to the Federal state unitary enterprise "Northern federal enterprise for radioactive waste management" of the Federal atomic energy agency"
- - R 2.6.6.57–04 "Radiation hygienic requirements for facilities of long-term storage of single reactor compartments of NS under dismantlement"
- - MU 2.6.6.22–05 "Radiation safety assurance during remediation of shore technical bases"

Despite the available legislative and regulatory basis in general allow to solve problems of radiation safety assurance in the course of FSUE "SevRAO" supervision, it is reasonable to develop and establish regulatory documents of the following content:

- Improvement of the system of medical response in case of radiological accidents, specification of a listing, order and time of performance of necessary emergency remedial technological operations
- Radiation safety assurance during design and performance of radiation-hazardous operations, application of personal, local and collective protective equipment
- Monitoring of radiation situation, occupational doses directly in the course of radiation-hazardous operation performance
- Environmental monitoring performance and filling up the database in the supervised areas of facilities
- Development of recommendations and criteria, hygienic regulations on remediation of contaminated sites, regulations of social acceptable guarantees of the public radiation safety during and after remediation of the site

Annex: Radiation Safety Regulation During Supervision at FSUE "SevRAO" Facilities

Aspects of radiation safety regulation	Regulatory documents with respect to radiation safety
Radiation hazard category of the facility; availability and borders of health protection zone and supervised area	OSPORB-99 SP HPZ and SA-07 «Health protection zones and supervised areas of radiation facilities. Operational conditions and justification of borders» MR 2.6.1.27-03 «Supervised area of radiation facility. Arrangement and performance of radiation monitoring»

(continued)

Annex: (continued)

Aspects of radiation safety regulation	Regulatory documents with respect to radiation safety
	MU 2.6.1.2005–05 «Establishment of potential hazard category of the radiation facility»
Availability of authorization to work using radiation sources (sanitary epidemiological certificates, licenses)	OSPORB-99
Subdivision of the industrial site area into some radiation hygienic zones according the levels of hazard	SP 2.6.1.2154–06 «Radiation safety assurance during overall dismantlement of nuclear submarines»
Measures for prevention of unauthorized access to RS, arrangement of sanitary warrant regime	par.10.17 SPP PUAP-03
	OSPORB-99
Performance of decontamination of surfaces of equipment and rooms	SanPiN 2.2.8.46–03 «Sanitary rules relating to decontamination of personal protective equipment»
	MUK 2.6.1.016–99 «Monitoring of superficial contamination of equipment, workshops, vehicles and other subjects with radionuclides»
Reference levels of radiation situation parameters	NRB-99, OSPORB–99
Observance of radiation safety rules at the facility, including in the course of radiation hazardous operations	OSPORB-99, NRB–99
Measures for prevention of radiological accidents, occupational, public and environmental protection in case of radiological accident	OSPORB-99
	R 2.6.1.47–01 «Standard contents of a plan of medical sanitary provision of the personnel and the public during radiological accidents»
	MU 2.1.6.30–04 «Arrangement and performance of radiation dosimetry monitoring in the course of mitigation of radiological accident consequences»
	MU 2.6.1.34–04 «Arrangement of emergency radiation monitoring of occupational external exposure during operations at radiation hazardous facilities of Minatom of Russia»
Subdivision of workers into A and B groups	OSPORB-99
Education and training of the personnel, permitted to work using RS	OSPORB-99
Arrangement of safe conditions of work using RS	OSPORB-99 and other scientific documents
	MU 2.6.1.04–05 «Radiation safety assurance during NS dismantlement»
	MU 2.6.6.22–05 «Radiation safety assurance in the course of remediation of shore technical bases»
	R 2.6.1.35–02 «Radiation safety assurance during IFA discharge of nuclear submarines under dismantlement» RBV-2002

(continued)

Annex: (continued)

Aspects of radiation safety regulation	Regulatory documents with respect to radiation safety
	R 2.6.6.37–02 «Hygienic regulations, established in the course of NS dismantlement»
	MU 2.2.8/2.6.167–02 «Arrangement of ventilation at radiation facilities»
	MU 2.6.1.44–04 «Health risk assessment for workers from the facilities involved in NS dismantlement»
RW management	SPORO-2002
	R2.6.1.42–02 «Radiation hygienic requirements for SRW allocation in reactor compartments of NS under dismantlement»
	RD 2.6.1.69–02 «Radiation monitoring of metal scrap, generated in the course of dismantlement of NS and above water ships equipped with NPI»
	MU 2.6.5.32–01«Radiation monitoring of metal scrap, generated in the course of dismantlement of NS»
Monitoring of radiation situation in workplaces, in workshops, on-site, in health protection zone and supervised area, that of discharge and effluent of radioactive substances	MU 2.6.1.14–01 «Radiation situation monitoring. General requirements»
	MU 2.2/2.6.1.20–04 «Assessment and classification of the personnel conditions of work at operations using RS»
	Methodic recommendations on sanitary monitoring of radioactive composition of environmental media, 1979
	Methodic guidance on detection and mitigation of radioactive contaminations, 1981
	SanPiN 2.3.2.1078–01 «Hygienic requirements for safety and food cost of foodstuffs»
	SanPiN 2.1.4.1074–01 «Drinking water. Hygienic requirements for quality of water from the centralized system of drinking water supply»
	MU 2.6.1.44–02 «Activity concentration of radionuclides in air of workplaces. Requirements for determination of annual activity value»
	MU 2.6.1.58–02 «Criteria and methods of assessment of conditions of radioactively contaminated lands, adjacent to the Navy bases»,
	MU 2.6.1.1868–04 «Introduction of radiation safety indexes with respect to the environmental media conditions, including food raw materials and foodstuffs, into the system of social hygienic monitoring»

(continued)

Annex: (continued)

Aspects of radiation safety regulation	Regulatory documents with respect to radiation safety
	MU 2.6.1.1981–05 «Radiation monitoring and hygienic assessment of sources of drinking water supply and drinking water by radiation safety indexes. Optimization of protective actions with respect to sources of drinking water supply with excess radionuclide contents»
	MU 2.6.1.11–06 «Radiation hygienic requirements for the system of radiation monitoring of facilities for long-term storage on the shore sites»
	MU 2.6.1.2153–06 «Operational assessment of the public doses under conditions of airbone radioactive contamination of the site»
Control and account of annual individual occupational doses of the personnel A and B groups, as well as excess authorized dose limits	OSPORB-99, NRB-99
	Guideline on arrangement and performance of personal dose monitoring No. 2925–83 Ministry of Health USSR
	R 2.6.1.17–03 «Guideline on investigation of unplanned or emergency occupational exposure of workers from Minatom of Russia facilities» RF Ministry of Health
	Methodic recommendations on performance of centralized personal dose monitoring of the occupational external exposure, 1995
	MU 2.6.1.25–00 «Dose monitoring of external occupational exposure» RF Ministry of Health
	MU 2.6.1.26–00 «Dose monitoring of internal occupational exposure»
	MU 2.6.1.05–01 «Recommendations on risk assessment of radionuclide exposure with the purpose of internal exposure regulation»
	MU 2.6.1.45–01 «Dosimetry. Determination of individual effective neutron doses»
	MU 2.6.1.56–02 «Monitoring of equivalent photon and beta doses in the skin and the lens of eye»
	MU 2.6.1.26–03 «Determination of external doses in case of radiological accident at Minatom's facilities» RF Ministry of Health
Arrangement of information system for personnel and the public with respect to radiation situation	OSPORB-99

(continued)

Annex: (continued)

Aspects of radiation safety regulation	Regulatory documents with respect to radiation safety
Quality of periodic medical examinations of workers	OSPORB-99 Orders of RF Ministry of Health No. 90, 83, 105
Annual presentation of radiation hygienic certificate of the facility, registration forms of the single state system of control and account of individual doses, ESKID	OSPORB-99
	NRB-99
Arrangement a system of physical protection at the facility	OSPORB-99
Contents and organizational structure of the Radiation Safety Service	

List of Abbreviations

NS	nuclear submarine
STB	shore technical base
LRW	liquid radioactive waste
SA	supervised area
RS	radiation source
SNF	spent nuclear fuel
RW	radioactive waste
HPZ	health protection zone
SRW	solid radioactive waste
FSUE "SevRAO"	Federal state unitary enterprise «Northern Federal enterprise for radioactive waste management»
NPI	nuclear powered installation

Regulative Provision of Waste Management Regulatory Supervision at SevRAO Facility

O.A. Kochetkov[1], S.G. Monastyrskaya[1], B.E. Serebryakov[1], N.P. Sajapin[1], V.G. Barchukov[1], and M.K. Sneve[2]

Abstract This paper summarises the regulatory framework in Russia for the management of radioactive wastes, with special emphasis on wastes arising in decommissioning work and on the distinctions between exempt waste, very low level radioactive waste and other radioactive waste.

Keywords Industrial waste management, radioactive waste (RW), decommissioning, man-made radionuclides, VLLW

Industrial waste management is one of the relevant challenges of the modern enterprise activity. It is relevant for nuclear facilities in particular. A large stream of wastes and materials is generated in the course of nuclear facility operation, and some of these wastes contains considerable amount of radioactive substances, forming Radioactive Waste (RW) group. A procedure of this waste management is currently known. The main criteria of waste categorization in terms of their specific activities had been established; principles and practices of their safe management had been specified.

During radioactive facility decommissioning, large streams of industrial wastes, containing man-made radionuclides with activity levels lower than those, which allow their ascription to RW category, are generated in addition to radioactive ones. Selection of economic and ecologically safe option of such material management is accompanied with some difficulties, because of their special features: very low radiation exposure to individuals at rather significant initial amounts.

Having in mind a high importance of this challenge, and with the purpose of the international consensus in approaches, criteria and digital parameters, needed for optimized decision making in all aspects o f waste management (containing man-made radionuclides with activity levels lower than those of their ascription to RW), IAEA implements systematic review and examination of the mentioned challenge. In 1988, based on examinations accomplished, IAEA issued the safety guidance No

[1]State Research Centre-Institute of Biophysics (IBPh), Russia
[2]Norwegian Radiation Protection Authority (NRPA), Norway

M.K. Sneve, M.F. Kiselev (eds.) *Challenges in Radiation Protection and Nuclear Safety Regulation of the Nuclear Legacy*,

89 «Principles for the Exemption of Radiation Sources and Practices from Regulatory Control. Safety Series», main provisions of which had been developed in the Safety Guidance «Application of the Concepts of Exclusion, Exemption and Clearance» RS-G-1.7 issued in 2006.

In the course of nuclear facility decommissioning (NPP, research reactors etc.), large waste streams are generated (soil, construction debris, metal). Generally, these waste contaminations with man-made radionuclides are a bit higher than exemption level from regulatory control, but lower than RW level. Therefore, separation of "very low level waste" (VLLW) category from the waste classification became reasonable. That is why, this category had been introduced into practice, and such wastes permitted to be disposed at the industrial waste landfills.

Now, there is no single opinion regarding VLLW management. In Sweden, the special landfill had been arranged for such waste disposal near NPP in Oscarshamn, where very low level wastes are disposed from all radiation hazardous facilities. In France, this waste is not conveyed to conventional waste disposal landfill, instead it is stored within industrial sites, where it was generated, or it is transported into RW disposal facility in La-Aube. In Japan, with the purpose of VLLW disposal resulted from the research JPDR reactor dismantlement, a special disposal facility had been built within the industrial site.

The solution of VLLW management, both in the world practice and in Russia is possible through introduction of this waste category into the system of the national classification. There is a framework, in Russia, which permits to solve the problem of such waste disposal (OSPORB-99 Para 3.11), but a lack of regulative and methodic documents hampers introduction of VLLW management system into everyday practice.

Comparative analysis of Russian and IAEA classifications (Table 1) shows that industrial waste category (materials and products) with low radionuclide contents corresponds in the international terminology to so-called category of Very Low Level Waste. With the purpose of the single interpretation of the waste categorization system it is reasonable to use the same term in Russian documents under development.

VLLW management problem arose in Russia critically in the course of nuclear legacy problem solving, especially during decommissioning and remediation of the former shore technical bases of the Navy (sites of spent nuclear fuel and radioactive waste temporary storage (SNF&RW STS)), including those located in Russian Northwest (Andreeva bay and Gremikha village). These waste amounts are comparable with RW amount accumulated and generated at SevRAO and DalRAO facilities, therefore the waste management problem is rather relevant.

Analysis of available (or possible) VLLW management technologies showed that the special features of the radiation hazardous facility must be taken into account when selecting the most safe and economically reasonable option of management. For example, legacy waste being generated during remediation of the former shore technical bases of the Navy differs considerably from those generated at NPP. These difference examples are as follows:

- At NPP, RW and industrial waste management is aimed at sorting and disposal of new generating wastes, while at SNF&RW STS this activity deals with legacy wastes.

Table 1 Comparative specification of radioactive substance containing wastes depending of disposal type

IAEA project N DS 390 - 2006			Type of disposal: IAEA	Type of disposal: Russia	Russian (OSPORB-99, SPORO-2002)		
RW	1	HLW with long-lived	Geological formations	Geological formations Underground	$T_{1/2} > 30$ years HLW	1	RW
	2	ILW with long-lived	Underground disposal	Underground	$T_{1/2} > 30$ years ILW	2	
				Shallow	$T_{1/2} < 30$ years (short-lived)		
	3	LLW with long-lived	Shallow		LLW	3	
	4	very short-lived $T_{1/2} \leq 100$ days up to EW levels (VSLW)	Storage for decay up to clearance levels, uncontrolled disposal		Short-lived waste $T_{1/2} < 15$ days up to $<$HAO $T_{1/2} < 1$ year up to RW SAMS	4	
Both RW and non-RW	5	Exemption level VLLW or a bit higher	Shallow disposal with limited control (landfills)	Disposal as conventional industrial wastes	Contaminated materials of limited use A_{sp} $<$LLW-0.3 (exemption level) Bq/g	5	Industrial waste
Non-RW	6	Exempt waste (EW)	Release from control, landfills, uncontrolled disposal		Contaminated materials of unlimited use $A_{sp} < 0.3$ Bq/g	6	

- An isotope compositions are rather different: NPP waste is characterized by low content of Sr^{90} and high content of short-lived isotopes, in particular Co^{60}; at SevRAO, the waste isotope composition consists mainly of Sr^{90} and Cs^{137}, where Sr^{90} quota is about 20%.
- Waste disposal facilities at NPP, are generally located within Health Protection Zones (PHZ), while at SevRAO they are arranged on the industrial site.
- Special features of SevRAO force to search for specific technologies of SNF&RW management, including VLLW.

In addition to differences in waste management technologies, radiation situation on the industrial sites is rather different too. At NPP, noticeable contamination of the site is practically absent, while at SNF&RW STS, in Andreeva bay in particular, according to NIKIET and SRC-IBPh experimental data, soil sample activities are in the range 10^2–10^4 Bq/kg, and at some parts of the site (near the brook in the vicinity of Building 5), specific activity of the soil reaches 10^5 Bq/kg.

It should also be noted that at the area adjacent to SevRAO, there are no settlements, agricultural grounds, forestry lands and water media, which the public could use to get foods.

At SNF&RW STS, VLLW management situation is complicated, because the design of these radiation hazardous facilities does not contain any waste management technologies for the post-operational period.

The data presented illustrates that the waste management problem solving at SevRAO facility requires elaborating provisions relating to VLLW management procedure, variants of release from regulatory control, disposal options, sitting the disposal landfills in terms of above mentioned features, forming both radiation situation and occupational doses.

The listed circumstances have defined the necessity of the regulative document development, which permits to perform efficient and effective radiation hygienic supervision of safe VLLW management assurance.

The main issues of the document are occupational, public and environmental protection. This document justifies the base radiation safety criteria – specific activity boundaries for VLLW category, which according to OSPORB-99 Para 3.11.4, are within the range 0.3–100 kBq/kg for β-emitting radionuclides. Having in mind the fact that β-emitting radionuclides Sr^{90} and Cs^{137}produce the main problem at SevRAO facility, the lower VLLW boundary (0.3 kBq/kg) could me determined more precisely.

Hygienic regulations "Contents of man-made radionuclides in metals" (GN 2.6.1.2159-07), approved by The Directive of the RF State Chief medical officer No 5 of 08 February 2007, recommend to release metals intended for re-use (with specific activity levels of 10 kBq/kg – for Sr^{90} and 1 kBq/kg – for Cs^{137}) from radiation control. In the light of mentioned SevRAO location and operation features, such exemption levels can be applicable to the waste disposed at SevRAO landfill, when building necessary engineering containment barriers, not affecting safety. In terms of the waste radionuclide composition on SevRAO industrial sites, and provided that inequation [1] is valid, metallic VLLW specific activity level authorized for exemption from regulatory control, will equal to 1.2 kBq/kg.

$$\sum_{i=1}^{N} \frac{A_i}{ПК_i} \leq 1, \quad [1],$$

where: N – the number of different radionuclides in the particular metal;
A_i – specific activity of i-th radionuclide in this metal, kBq/kg;
$ПК_i$ permissible specific activity value of i-th radionuclide in this metal, kBq/kg.
Taking into account 100 year planned duration of control over the landfill, as well as 30 years Sr^{90} and Cs^{137} half-lives, we may emplace wastes with 12 kBq/kg average specific activity.

At that, the containment barriers must assure non-exceeding of intervention levels by water over this period, because other pathways of radionuclide migration from the landfill are irrelevant.

In addition to selection and justification of safe VLLW management, development of requirements for safety assurance during the waste disposal is rather important. Here, VLLW disposal safety will depend upon:

- The waste activity concentration on the landfill during a year («radionuclide capacity» of the landfill)
- Radionuclide composition of VLLW
- Containment barriers
- Minimization of radionuclide intakes from VLLW into the environment
- Conditions of the landfill release from radiation control
- Prohibition of fire-, explosive-, and chemical hazardous substance disposal

The landfill safety must also be provided during its release from radiation control. The extent of regulatory requirements at that level of the landfill operation is determined by scenarios of possible further use both of the site and the landfill. For VLLW disposal landfill at SNF&RW STS in Andreeva bay, the following scenarios are possible:

- Exemption from regulatory control provided that the specific activity being averaged by the whole landfill, does not exceed 0.3 kBq/kg («greenfield»)
- Limited use of the waste disposed with activity level higher than 0.3 kBq/kg
- "Brown lawn" arrangement on-site STS keeping, at the same time, the landfill under conservation there and transferring the industrial site area into the "federal reserve lands" category

When selecting the most acceptable scenario, safety justification and coordination with the supervision bodies is necessary. In the course of decision making, one must recognize that effective dose for the critical group of population, due to the residual contamination, must not exceed 0.1 mSv/year, according to IAEA recommendations and in compliance with the Guidance R 2.6.1.25 – 07 «Remediation criteria and regulations of sites and facilities from the federal state unitary enterprise "Northern Federal Facility for radioactive waste management (SevRAO)" of the Federal Atomic Energy Agency, contaminated with man-made radionuclides».

Summarizing all above mentioned, we can conclude that:

1. Russian sanitary legislation envisages general system of the waste management, which waste corresponds to VLLW (very low level waste) international category.
2. To harmonize terminology of Russian regulative basis with international one, industrial waste containing man-made radionuclides with specific activity levels lower than LLW, but higher the level of exemption from radiation control, are reasonable to separate into VLLW category.
3. Having in mind planned operational conditions of SevRAO VLLW landfill: arrangement of containment barriers, 100-year duration of radiation control of the landfill conditions, and large distance from any settlements etc., we consider that 30 kBq/kg waste is suitable for convey to disposal at the landfill.

Regulatory Case Studies and Western Experience Concerning Nuclear Legacy

Site experience from Centre de Cadarache, CEA, France

C. Deregel and F. Gauthier

Abstract This document presents a summary of on-going activities for the recovery of radiological waste stored in trenches in the 1970s on the CEA Nuclear Research Centre of Cadarache (France).

After an introduction explaining the origin of the waste, the activities to be performed for its recovery and the necessary installations are described; then the final destination of the recovered waste is presented (storage, disposal, treatment and, when possible, recycling).

This document should be read together with the document "Issues in Decommissioning and Remediation of Nuclear Legacy Sites" written by the same author.

Keywords Recycling, ANDRA, decommissioning, remediation

1 Foreword

The Nuclear Research Centre of Cadarache is operated by the French Nuclear Energy Agency (CEA) and has been created in 1959.

The operation of several research reactors and laboratories produces radioactive waste which was at the very beginning temporary stored on the site in a dedicated facility built in the 1960s.

In 1969, the CEA decided to experiment the storage of low level radioactive waste in trenches in order to study the behaviour of the waste and the transfers of radionuclides in the soil.

This experiment has been declared to the IAEA.

On the administrative point of view, the trenches area is part of a storage facility classified as "Nuclear Basic Installation" (INB no. 56), authorized by the French

Institut de Radioprotection et de Sûreté Nucléaire (IRSN), France
Contributions of CEA Cadarache Research Centre

M.K. Sneve, M.F. Kiselev (eds.) *Challenges in Radiation Protection and Nuclear Safety Regulation of the Nuclear Legacy*,

Nuclear safety Authority and operated according to the regulatory framework in force for INBs.

From 1969 up to 1974, 3,000 m^3 of low level radioactive waste have been buried in five trenches, some of them conditioned in drums or in vinyl bags, other ones not conditioned (mainly metallic pieces and rubble).

In 1995, a first limited operation of recovery of waste was performed in order to evaluate the behaviour of the waste and the transfer of contamination in soil and to prepare the remediation of the trenches.

The lessons learnt have been used for the planning of the recovery of the whole waste stored in trenches of the INB 56 and its final disposal in existing French repositories and, for the categories of waste not accepted in the repositories, its temporary storage in the new facility "CEDRA" which was put in operation in Cadarache in 2006 (a new INB).

The operations of recovery of waste stored in the trenches started in 2004 and are still in progress.

2 Short Description of the Activities and Installations

2.1 Activities and Requested Installations

The remediation of one trench requests following activities with constraints of containment and radiological protection:

1. Prevention of the risk of dissemination of contamination in the environment (Installation of a shelter over the trench)
2. Withdrawal of the soil covering the trench
3. Extraction of the waste from the trench without contamination of the shelter (a mobile extraction cell and a tarpaulin assuring the containment of the trench including the extraction zone)
4. First characterisation and sorting of the waste inside the extraction cell
5. Transfer of the waste up to an area for temporary storage and distribution to the different working stations (a trolley and a distribution cell)
6. Inside the distribution cell, gamma spectrometry and counting for identification of radionuclides
7. Transfer to working stations for opening of the packages, sorting, new packaging of the waste when necessary, identification of the new packages (activity, radionuclide inventory, final conditioning (mainly in 100 l drums) and identification of the final packaging according to acceptance criteria for the next destination of the waste (Surface site disposal facilities operated by ANDRA or temporary storage in the newly built storage facility CEDRA, in function of the radiological waste package inventory)

2.2 Short Description of the Facilities Used

Besides the buildings supporting the technical activities for the recovering of waste (offices, workshops, buffer storage), the specific equipments developed for the operation are described hereafter.

2.2.1 Shelter

The shelter is a metallic hangar which can be dismantled covering one trench. It contains all technical equipments necessary for the recovery of waste. Supporting equipments (ventilation, sanitary pass, cloak-rooms) are attached to the shelter.

2.2.2 Extraction Cell

The extraction cell is a mobile structure supported by rails installed on the two side of the trench; it contains hoisting and grapping devices for the recovery of items which were buried in the trench; this installation is under dynamic containment, the containment being ensured by a tarpaulin covering the trench and the extraction cell.

Lock chambers allow the access inside the cell for the personnel and for the trolley used for the confined transportation of waste inside the shelter.

2.2.3 Distribution Cell

The distribution cell is a fixed structure installed inside the shelter. It receives the trolley carrying the waste extracted through a lock chamber and contains the equipments for handling, cutting if necessary, sorting and conditioning the waste before its management in the characterisation cell.

2.2.4 Characterisation Cell

The characterisation cell is a fixed structure installed inside the shelter and connected to the distribution cell. It contains the equipments for detailed sorting and final conditioning of waste (glove boxes), characterisation (spectrometers, gamma scanning) and identification of the packages and an air-lock for the going out of the finally conditioned waste packages.

2.2.5 Trolley

The trolley is a self propelled vehicle running on rails allowing the confined transport of waste between the extraction cell and the distribution cell.

3 Existing French Facilities for Disposal and Treatment of Low and Intermediate Level Short Lived Radioactive Waste

3.1 *Waste Disposal Facilities*

The French Agency for radioactive waste ANDRA operates two disposal facilities:

- The "Centre de Stockage de l'Aube" for low and intermediate level short lived waste (surface repository in concrete vaults)
- The "Centre de Stockage TFA" for very low level waste (surface repository in trenches digged in a clay medium)

3.2 *Treatment Facilities*

The French Society for Industrial Waste Treatment SOCODEI operates the basic nuclear installation CENTRACO (Nuclear Waste Treatment and Conditioning Plant) located near the Marcoule site (Rhone valley). CENTRACO contains, on a single site, an incineration unit for burning liquid and solid waste, a melting facility for metallic waste and a recycling unit for scrap metal.

About Activity of the Federal Medical Biological Agency in the Field of the State Safety Regulation at Atomic Energy Use

V.V. Romanov

Abstract This paper describes the role of the FMBA in supervision of radiation safety in Russia and its links with other regulatory bodies. It sets out the FMBAs special activities related to nuclear legacy management and defines a list of important issues with need to be addressed in the short and long term.

Keywords Federal medical-biological agency (FMBA of Russia), sanitary epidemiological supervision, OSPORB-99, NRB-99

According to the current legislation of the Russian Federation, the Federal medical-biological agency (FMBA of Russia) is now responsible for medical sanitary provision and for the state sanitary epidemiological supervision in organizations of some industrial branches where conditions of work are extremely hazardous, as well as for such supervision of the population of some areas, according to the listing approved by the RF Government. In addition, FMBA of Russia is responsible for the state safety regulation at atomic energy use.

FMBA of Russia operates directly and through its territorial supervision bodies and lower organizations.

FMBA's structure (Fig. 1) includes The State sanitary epidemiological supervision Department of the central FMBA of Russia office, 42 regional (interregional) departments, 63 centers of hygiene and epidemiology, 18 research institutes, 91 patient care and prophylactic institutions arranged in compliance with the nearest approach to the subject of supervision.

At that, FMBA territorial bodies and centers of hygiene and epidemiology are enclosed in the single centralized system of bodies and authorities responsible for the state sanitary epidemiological supervision in the Russian Federation. This system works in compliance with the Directive of RF Government No 569 of 15 September 2005 "Provisions of implementation of the state sanitary epidemiological supervision in Russian Federation".

Deputy Head of the Federal Medical Biological Agency (FMBA), Russia

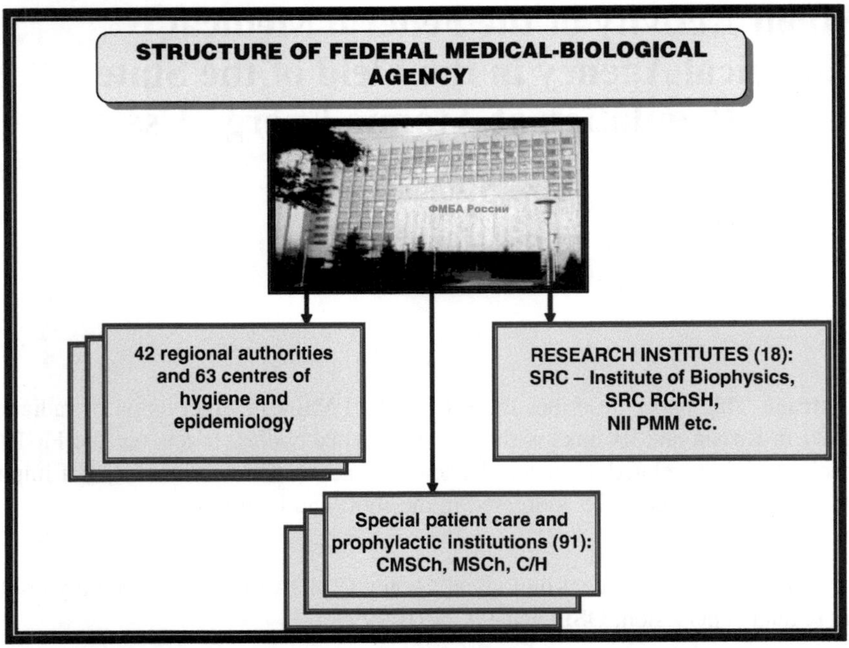

Fig. 1 Structure of the Federal medical-biological agency

The FMBA's research institutes implement scientific support for activity of practical health institutions, territorial bodies, centers of hygiene and epidemiology of FMBA of Russia, including, health examination of the contingent under service and provision of sanitary and epidemiological supervision in the field of elaboration of regulative legal acts relating to control of radiation, chemical and biological hazardous organizations.

Within the Federal target program "Nuclear and radiation safety of Russia" for 2000–2006 relating to the section V «Regulative support for the state regulation of nuclear and radiation safety», sub-section 19 «Elaboration of federal norms and rules of radiation safety» (sanitary hygienic aspects), FMBA's research institutes developed 18 sanitary regulations (all of them had been registered in RF Ministry of Justice), more than 100 guidance documents. Guideline, and methodic documents for measurement performance and introduced them in its bodies of the state sanitary epidemiological supervision as well as in its lower organizations.

Available legal and regulatory basis permits in general to solve problems relating to radiation safety assurance during special and hazardous operations, improves main legal documents, including NRB-99.

Annual analysis both of radiation certificates of the Russian territories, and reports of FMBA territorial bodies shows that radiation situation is stable and remains satisfactory in general. Radiation factor is not a leading factor of unhealthy impact neither on the personnel of radiation facilities, nor the public living in the supervised areas.

During last five years, any cases were registered of exceeding of radionuclide effluent/discharge limits; chemical contaminations of common air and water media do not exceed maximum permissible concentrations.

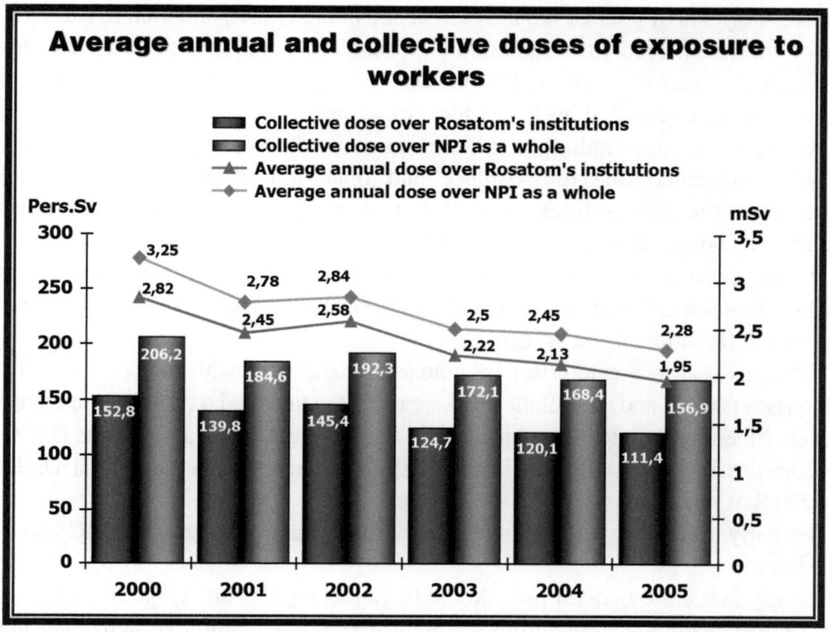

Fig. 2 Average annual and collective doses of exposure to workers

The state of radiation safety at Rosatom's facilities is assessed as satisfactory.

According to data of many-year monitoring, annual occupational dose within the branch is not more than 2 mSv (Fig. 2).

Indexes of chronic occupational morbidity at facilities under FMBA service, during recent years are lower than in the Russian Federation in general (Fig. 3). Incidence of acute occupational pathology occurs very rare, mainly, due to emergencies.

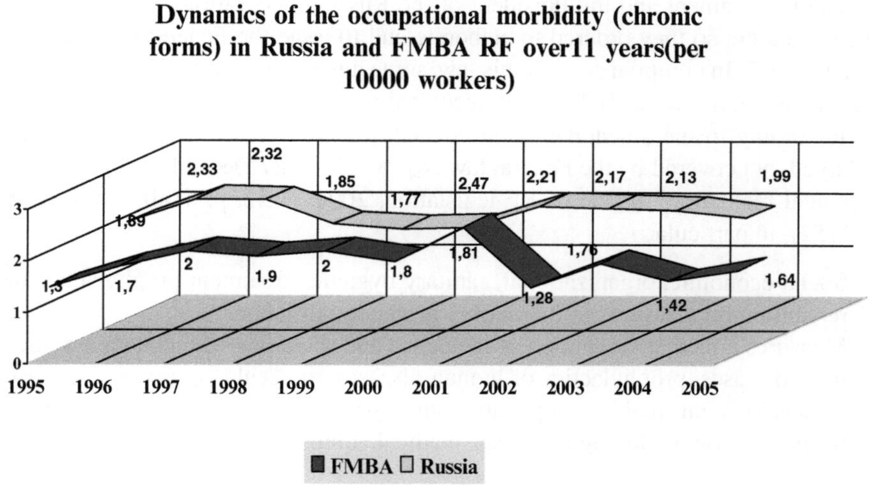

Fig. 3 Dynamics of the occupational morbidity in Russia

Over 60 years of nuclear industry existence in the USSR, and then in Russia, 753 persons were suffered from radiation exposure, including 349 persons with registered acute radiation sickness; the number radiation-induced deaths is 71 (taking account of the Chernobyl NPP accident consequences).

At the same time, radiation situation in some Russian regions is assessed as rather complicated due to existence and operation of nuclear and radiation hazardous sites and facilities, as well as to consequences of their operation and occurred accidents and abnormal situations.

At the present time, 4.3 millions cubic meters of liquid radioactive waste, 36,000 t of solid radioactive waste and 574,000 t of non-radioactive waste are accumulated at facilities of the branch.

Sixteen Rosatom's enterprises have areas contaminated with radionuclides. Three enterprises possess and control the most areas of contaminated territories, these are: PA Mayak, Siberian Chemical Combine (SCC) and Mining Chemical Combine (MCC).

Complex radiation situation is observed in the area of PA Mayak, at DalRAO and SevRAO facilities.

Recently, a number of challenges have arisen due to change of the RF existing legislation.

In July 2003, the Federal law "About technical regulation" came into force; this law defined technical regulations as main and single regulative and legal documents, establishing compulsory requirements for products, services, operations, processes and other subjects of technical regulation.

The Government planned that some elaborated and approved generic and special technical regulations replace different instructions, state standards and norms, and these regulations have status of either federal laws or the Governmental directives. Therefore, in 2004, A Plan of technical regulation elaboration was approved for the next two years. In 2006, this Plan was amended and supplemented.

Up to now, the RF Government has approved the only special technical regulation "About requirements for effluents of hazardous substances (contaminants) from machinery issued within the Russian Federation territory".

The Government and the President of the Russian Federation could not ignore such situation, so they ordered to elaborate and to issue the Federal law No 65 of 1 June, 2007. In compliance with this law, some amendments have been introduced into the Federal law No 184 «About technical regulation».

According to the amendments introduced, a listing of actions is clarified and enlarged, not covered by the Federal law No 184-FZ of 27 December, 2002 "About technical regulation" (Code of RF legislation, 2002, N 52, p. 5140; 2005, N 19, p. 1752), in particular:

- Social economic, organizational, sanitary hygienic, treatment prophylactic, and rehabilitation measures in the field of protection of labour
- Measures aimed at prevention of incidence and spreading of human mass infectious diseases, prophylactics of human diseases, medical care (except for cases of development, making, application and execution of obligatory requirements for production, including remedies, medical equipment, foodstuffs)

- Measures directed to preservation of soil, common air, water resort media, water media considered as places of tourism and common rest (amendments of Article 1 of the Federal law N 184-FZ "About technical regulation")

The law postulates that the technical regulation can by established by the federal law or the directive of the RF Government; a listing is also given enclosing 17 urgent technical regulations, which must be adopted until 1 January 2010.

Moreover, cardinal changes had been introduced into Article 5 of the Federal law N 184-FZ "About technical regulation", according to which compulsory requirements for Rosatom organizations, in addition to requirements of the technical regulations are those established by the federal authorities, responsible for the state safety regulation at atomic energy use.

Another problem, which in the nearest future can affect safety of nuclear energy use facilities, deals with the fact that 18 December 2006, the RF President signed the Federal law No. 232 «About introduction of amendments into the Town-planning codex and some legislative acts of the Russian Federation».

This Federal law came into effect since 1 January 2007 and changed cardinally activity of the RF state sanitary epidemiological supervision bodies, including Federal medical-biological agency and its territorial bodies in the field of arrangement and implementation of preventive sanitary supervision. Due to these changes, all mentioned authorities became *per se* debarring from participation in consideration of construction designs, including those of radiation and chemical-hazardous manufactures, limiting their functions by the stage of sitting.

At the same time, the RF Government adopted the Federal target program of nuclear power energy combine development in the State, which, program, envisages speeded construction of nuclear power station compartments, increasing means for radioactive waste treatment, building of floating nuclear stations in Severodvinsk and other RF regions, enhancement, upgrading and re-equipment of nuclear fuel cycle enterprises.

The scope of FMBA problems to be solved in the nearest time and in future, is rather large and includes a number of directions, approved by the RF Government in the Federal target program "Nuclear and radiation safety assurance for 2008 and for the period up to 2015". These directions are as follows:

- Reconstruction of available industrial buildings for manufacturing of anti-radiation remedies and substations of remedies for prophylactics and treatment of radiation injures
- Arrangement and introduction of medical dosimetry registers of workers from nuclear and radiation-hazardous facilities
- Arrangement and ensuring of escalation and keeping preparedness of medical sanitary teams and regional centers
- Purchase of up-today high-sensitive instrumentation for dosimetry and radiometry control of different radiations in the environment and workshops of radiation hazardous facilities, as well as spectrometers for determination of human internal doses due to radionuclide intakes

- Capital construction, including construction of the radiological building and reconstruction of clinic No&$$$; 6 divisions for reception and treatment patients with acute radiation sickness being exposed in the course of emergencies and abnormal situations
- Construction of the medical centre for persons exposed to external or internal radiation, on the base of the federal state science institution of the Ural science and practical centre of radiation medicine

At the same time, according to international requirements, the following problems are to be solved:

- Regulation of human radiation exposure
- Prediction of development of heritage and cancer consequences for persons exposed
- Radiation situation monitoring and control in radiation hazardous facility locations
- Interdepartmental cooperation during operations aimed at radiological accident prevention and mitigation of their consequences
- As well as a number of organizational problems

Therefore, FMBA of Russia faces with the following urgent challenges:

1. To introduce amendments and supplements into The Provisions about the Federal medical-biological agency, approved by the RF Government Directive No 206 of 11 April 2005, assigning by this FMBA functions relating to the state safety regulation during nuclear energy use (direct Governmental order)
2. To take active participation in development of generic and special technical regulations in the field of nuclear energy use
3. To made a Convention with Rostechnadzor:
   (a) About a procedure of sanitary epidemiological review implementation with respect to construction of sites of nuclear energy use and participation of FMBA specialists in this work
   (b) About elaboration of a joint document defining special features of technical regulation in the field nuclear energy use
4. To revise and re-publish (introduce amendments and supplements) the existing regulatory and legal acts (sanitary rules) in the field of radiation safety, published over the period 1999–2006, including OSPORB-99 and NRB-99
5. To complete work with regulatory and legal acts planned to be approved in 2007; to issue a new edition of Sanitary Rules for design and operation of nuclear stations, hygienic requirements for radiation safety assurance at facilities of OAO "TVEL" involved in mining and enrichment of uranium ore, regulatory and legal acts with respect to radiation safety assurance at SevRAO facilities
6. To complete work connected with amendment and supplement of the listing of organizations under service of the Federal medical-biological agency, approved by the RF Government Directive No&$$$; 1156-r of 21 August, 2006 (in edition of the RF Government Directive of 16 December 2006 No 1745–r), to assign

functions of the state sanitary epidemiological supervision of FMBA to the following Rosatom organizations:

(a) OAO «Khiagda» Bagdarin village, Baunstovsky region, Buryatia republic
(b) OAO «Dalur» Ouksyanovskoe village, Dalmatovsky region, Kurgan district
(c) OAO "Novosibirsk plant of chemical concentrates" Novosibirsk city
(d) FSUE PA «Sever» Novosibirsk city etc.

7. To ensure preparedness of territorial bodies and centers of hygiene and epidemiology – together with medical teams of CMSCh/MSCh, C/H, as well as EMRDC of FSUE SRC-IBPh – for work under conditions of possible radiological accidents
8. To ensure compulsory participation of FMBA territorial bodies, jointly with the FMBA Department of the state sanitary and epidemiological supervision:

(a) In the procedure of sitting for construction of new power units, reconstruction and upgrading of NFC enterprises
(b) In the public hearing at the federal and local levels relating to construction of new power compartments and prolongation of the existing power units life cycles
(c) Jointly with FMBA patient care and prophylactics institutions – in assessment of the public health living within the supervised areas of the existing NPP and in regions of assumed construction, as well as examination of the habitat conditions (zero background)
(d) In coordination of different types of operational documents, entering to the territorial bodies and department of the state sanitary and epidemiological supervision for review, providing, at the same time, feedback the department of the state sanitary epidemiological supervision – territorial body

Radiation Protection of the Public and Environment Near Location of SevRAO Facilities

N.K. Shandala[1], A.V. Titov[1], N.Ya Novikova[1], V.A. Seregin[1], M.K. Sneve[2], and G.M. Smith[3]

Abstract This work presents the results of original investigations of radiation-hygienic situation, existing near location of SevRAO Facilities – STS of SNF and RW in Andreeva bay and Gremikha village. The obtained data permits to conclude that hard and long-term remedial work will be initiated after SNF removal. Having in mind modern approaches to guaranteeing radiation safety, the primary attention at remediation scenarios development was paid to justification of dose constraints of the residual contamination exposure to workers and the public. Four principal options of STS remediation were considered – renovation, conversion, conservation, and liquidation. Primary and derived quantitative radiation-hygienic criteria were formulated for each option.

Keywords Site of temporary storage (STS), spent nuclear fuel (SNF), radioactive waste (RW), the public, environmental media, monitoring, remediation

1 Definition and General Provisions

According to Federal law of Russian Federation "About radiation safety of the public", "radiation safety of the public" means a protectability conditions for actual and future human generations against ionizing radiation exposure affected their health.

According to Federal law of Russian Federation "About preservation of the environment", "environmental safety" means protectability of both environment and vitally important interests of human against possible negative impact of economic and any other types of activity, emergencies of natural and man-made origin, and their consequences.

Development of actions directed to the public protection and preservation of the environment against possible unhealthy radiation exposure of the STS facilities should be carried out at the projecting stage and at each stage of the radiation sources management, in terms of the radiation protection principles (dose limits, justification and optimization).

[1]State Research Centre – Institute of Biophysics (IBPh), Moscow, Russia
[2]Norwegian Radiation Protection Authority (NRPA), Norway
[3]Enviros Consulting Ltd., UK

With the purpose of routine STS operation, the following classes of standards are established:

- The principal dose limits
- Acceptable exposure levels, derived from the principal dose limits
- Control levels taking into account the STS facility safety level practically achieved

2 Characterization of SevRAO Facilities in Andreeva Bay and Gremikha

The STS Andreeva Bay is located on Kola Peninsula in the Barents Sea coastal strip (Motovsky gulf, west bank of Zapadnaya Litsa bay). The nearby settlements are: Bolshaya Lopatka (2.4 km); Nerpitchie village (1.8 km); Zaozersk city (8 km). The population is 15,700, the majority of which are military estates. The facility holds about $1.3 \cdot 10^{17}$ Bq of SNF and $6.0 \cdot 10^{14}$ Bq of RW.

The STS Gremikha is located on Kola Peninsula in Chervyanaya Bay of the Barents Sea. The nearby settlements are: Gremikha village (0.7 km from the site) and Ostrovnoy city (1.2 km). The population is 3,500 (mainly, former soldieries and their families). The facility holds about $1.3 \cdot 10^{16}$ Bq of SNF and about $3.3 \cdot 10^{13}$ Bq of RW.

Up to now, a large amount of SNF contained in 88 unloaded cores, as well as 17,558 t of solid radioactive wastes and 3,042 t of liquid radioactive wastes have been accumulated in Andreeva Bay and Gremikha.

After termination of operations in the 1980s, the infrastructure of the bases degraded and poor conditions of some building constructions led radioactive contamination of some parts of STS territory. It is evident that the process of SNF and RW management and remediation of STS territories will take many years and will require, apart from developing a new infrastructure, the efficient control over observance of radiation protection requirements.

3 Specification of Areas Within the STS Territory

With the purpose of radiation protection of workers and the public, the following areas are specified on-site and around the STS site (see under the STS plan):

- Controlled access area (CAA) – SNF and RW store facilities are situated here and radiation-hazardous operations are performed here too.
- Uncontrolled (free access) area (UA) – Facilities intended for work supplying in CAA.
- Health protection zone (HPZ) – This is an area of administrative and technical provision of the STS.
- Supervised area (SA) – This is an area surrounding the STS, where radiological monitoring is carried out to guarantee radiation safety and protection for the public.

The member of the public must not stay within the first three areas.

4 Radiation Situation On-Site the STS in Andreeva Bay

The accomplished examinations showed that gamma dose rates within the STS territory varied over a wide range: in CAA – from 0.2 to 140 μSv/h^{-1}; in UA – from 0.2 to 12 kBq·kg^{-1}; in HPZ – from 0.1 to 0.2 kBq·kg^{-1}. Within SA, gamma dose rates varies from 0.063 to 0.14 μSv/h^{-1} with an average value of 0.12 μSv/h^{-1}, which does not differ from the levels typical for the territories of Northwest Russia and in the Murmansk region, in particular. The results of selective personal dose monitoring show that external exposure gamma rates of the public and workers of group B (individuals who are not working directly with the sources of ionizing radiation, but who, due to their working place location, can be exposed to radiation) due to natural and man-made sources of ionizing radiation are, respectively, equal to 0.8 and 0.9 mSv/y. Internal public radiation doses associated with intake of radionuclides with food are 14 μSv·year^{-1}. The total effective radiation doses to the public living in the STS's SA of Andreeva Bay(due to natural and man-made radionuclides) are estimated to be approximately 0.8–0.9 mSv·year^{-1}, that is not more than the actual norms.

The highest level radioactive contamination of soil on-site induced by man-made radionuclides is observed in the area of the old technological pier and around some SNF store facilities, where ^{137}Cs specific activity reaches 5.7 10^{7} Bq·kg^{-1}, and that of ^{90}Sr is 5.7 10^{6} Bq·kg^{-1}. ^{137}Cs and ^{90}Sr concentrations in soil within HPZ and SA is at the background level typical for "clean" Russian Northern areas and does not exceed 36 and 4 Bq/kg, respectively (Fig. 1).

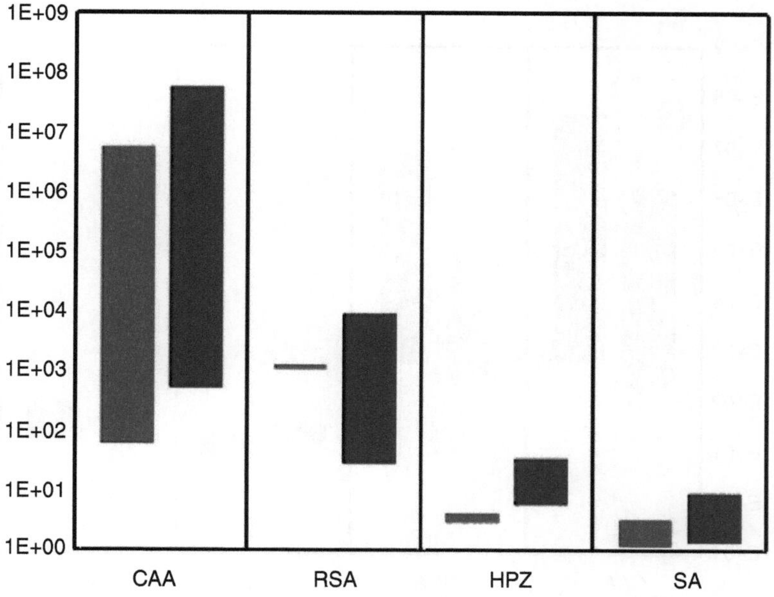

Fig. 1 ^{90}Sr and ^{137}Cs contents in soil on STS in Andreeva Bay

5 Radiation Situation On-Site the STS in Gremikha

Gamma dose rate within CAA varies from 0.2 to 500 μSv·h^{-1} (maximum values are four times more than those in Andreeva bay); in UA – from 0.2 to 12 μSv·h^{-1} and levels within approximately 80% of the territory do not exceed 5 μSv·h^{-1}. In HPZ and SA (in Ostrovnoy and Gremikha) it varies from 0.09 to 0.2 μSv·h^{-1}, i.e., within fluctuation limits of natural background of this region. The results of selective personal monitoring of the people living and working (workers group B) due to natural and man-made sources of ionizing radiation in the STS area show that the external exposure gamma dose rates are 0.7 mSv·year^{-1} (for public) and 0.9 mSv·year^{-1} (for worker group B). Internal public radiation doses due to intake of ^{137}Cs and ^{90}Sr with food are approximately 14 μSv·year^{-1}, which is significantly lower than acceptable levels.

Within the industrial site, man-made contamination is observed in top-soil due to ^{137}Cs, ^{90}Sr and, in small concentrations, ^{60}Co, ^{152}Eu, and ^{154}Eu. In SA (including Gremikha and Ostrovnoy),^{137}Cs and ^{90}Sr contents in soil are mainly within background level (1–50 Bq·kg^{-1}). In some cases, at local parts outside the settlements, observed levels exceed background values by up to 100 Bq·kg^{-1} by ^{137}Cs (Fig. 2).

6 Results of Radiation Situation Assessment at STS in Andreeva Bay and Gremikha

The results of neutron radiation rate measurements in SNF location areas showed that neutron radiation exposure of workers is not significantly high. The neutron radiation levels are close to the background values of 10^{-2}–10^{-3} μSv·h^{-1} (dose rate

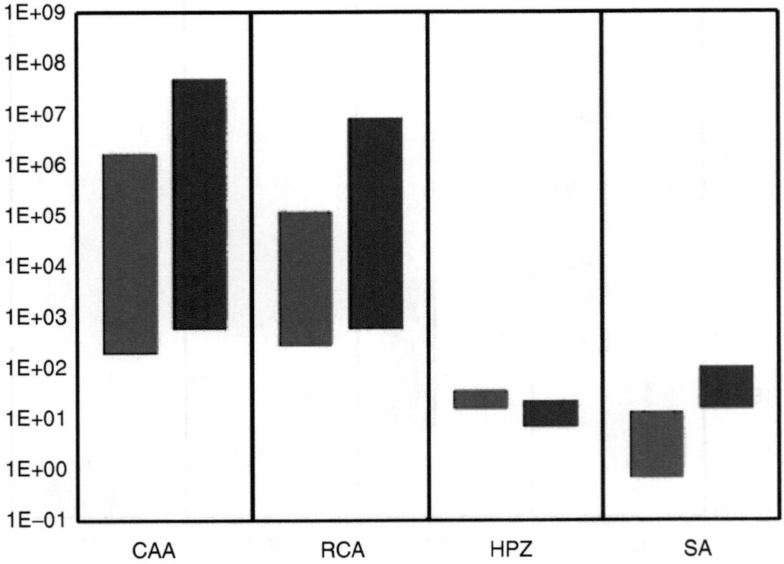

Fig. 2 ^{90}Sr and ^{137}Cs contents in soil on STS in Gremikha

of neutron component of cosmic radiation at the sea level is about $9 \cdot 10^{-3}$ µSv·h^{-1}), while the intensity ratio of gamma radiation to neutron radiation reaches 10^4–10^6.

Radiation monitoring of the environmental media showed considerable exceeding of typical background values of ^{137}Cs and ^{90}Sr radionuclide concentrations (in the SSZ coastal strip) in seaweeds, bottom sediments and vegetation. An exceeding is also observed in some cases in the STS SA environmental media in comparison with background values. Preliminary results of sorption experiments of radionuclides on local soil and ground waters suggest that radionuclide migration from highly contaminated areas on site, via groundwater flow pathways, is possible. This leads to permanent entry of radioactive substances into the off-shore marine environment.

According to radiation monitoring of catches in the STS off-shore marine environment, the concentration of ^{90}Sr and ^{137}Cs in fish is in the range 0.7–13 Bq·kg^{-1} for ^{90}Sr and 0.4–35 Bq·kg^{-1} for ^{137}Cs, respectively, being significantly lower than actual Russian accepted radiation contamination levels. With the purpose of radiation exposure restriction during large-scale STS remedial work, Federal Medical-Biological Agency established a public radiation dose quota; this quota is 100 µSv·year^{-1} due to effluents and 30 µSv·year^{-1} due to radioactive substance discharges (Table 1).

7 Remediation of Contaminated Territories

The analysis of radiation-hygienic situation taking account of the data obtained, leads to the conclusion that a long and extensive rehabilitation program is necessary after the removal of SNF and RW from the STS's. Having in mind up-to-date approaches to radiation protection, the attention, in identification of remediation scenarios, has focused on justification of reduction of the occupational and public exposure arising from of residual contamination.

8 Three Main Variants of the STS Remediation

1. Conservation (storage under surveillance) – excludes the potential threat of contamination of the STS territory, water area and air media. A guarded area is arranged and continuous radiation monitoring is carried out.

Table 1 Public radiation dose quota due to effluents and to radioactive substance discharges under conditions of SevRAO facility normal operation

Sources of exposure	Quota (µSv·year^{-1})
Gas-aerosol discharges	100
Intake with seafood	30
Reserve for unregistered sources	20

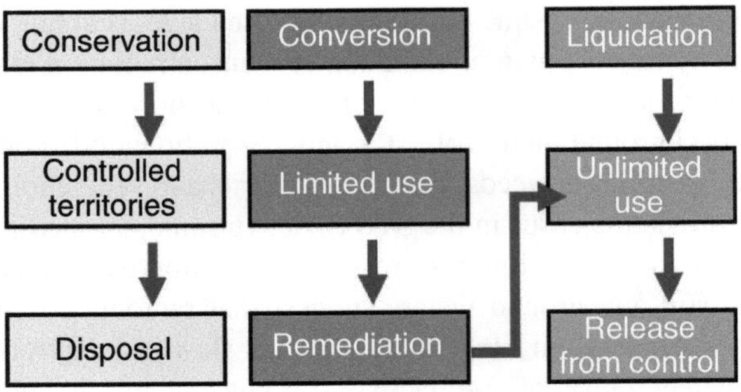

Fig. 3 Three main variants of the STS remediation

2. Conversion (renovation) – suggests subsequent use of the STS territories and facilities in compliance with the existing regulatory documents regulating the radiation impact on personnel and public under normal conditions of operation with radioactive sources. Limited use of the territory in combination with rehabilitation measures and radiation monitoring ("brown field" concept) is envisaged.
3. Liquidation – suggests stage-by-stage dismantling and removal of equipment, removal of RW, including contaminated environmental objects, and guarantees of limited exposure dose for critical group of public at the level $1\,mSv{\cdot}year^{-1}$ ("Greenfield" or unlimited use concept) (Fig. 3).

The STS in Andreeva Bay is not likely to be used for direct purpose in the future. The planned operations are associated with preparations and removal of SNF and RW from the territory with subsequent liquidation or conservation of the buildings and other constructions, and decontamination of the territory.

We assume that at STS in Gremikha, apart from environmental rehabilitation operations, the remediation and reconstruction of the infrastructure for unloading and following interim storage of NS core reactors with liquid-metal coolant, is required.

9 Norms of Remediation

Federal Medical-Biological Agency approved norms for the main variants of the STS remediation (Table 2). They had been developed on the base of actual Russian laws and standards and taking account of radiation situation existing currently at the STS. These norms had been developed in terms of the contemporary international recommendations and experience in the field of contaminated area remediation in other States.

Table 2 Dose constraints for different options of the STS remediation

Option of remediation	Individual category	Dose constraint (mSv/year) Due to the residual contamination	Due to new activity	Total	Dose limit
Conservation	Workers	2	–	2	20
	The public (within SA)	0.1	–	0.1	1
Conversion	Workers from A group	3	7	10	20
	Workers from B group	1	1	2	5
	The public (within SA)	0.1	0.15	0.25	1
Liquidation	The public (within the former STS area)	1	–	1	–
	The public (the rest area)	0.1	–	0.1	–

10 Conclusion

The described work carried out under joint Federal Medical-Biological Agency and Norwegian Radiation Protection Authority Project, devoted to regulation of the public radiation and nuclear safety during SevRAO operations, current output has included the following documents:

- InitialThreat Assessment for the situation at SevRAO sites
- Guidance "Criteria and norms on remediation of SevRAO sites and facilities contaminated with man-made radionuclides"
- Guidance "Hygienic requirements for personnel and public radiation safety guaranteeing at the stage of designing the work with SNF and RW at FSUE "SevRAO" Branch No 1"

The work completed and in progress will allow many problems to be solved. However, as the engineering project plans are developed in more concrete detail, additional data on the site conditions will become available, and prognostic assessments will become more accurate.

Special Features of the Personnel Radiation Protection Assurance During SNF and RW Management at SevRAO Facilities

A.V. Simakov[1], O.A. Kochetkov[1], Yu.V. Abramov[1], M.K. Sneve[2], and A.V. Grigoriev[3]

Abstract Today, overall decommissioning of nuclear submarines (NS) belonging to the Northern Fleet RF and remediation of available sites of temporary storage (STS) of the spent nuclear fuel (SNF) and radioactive waste (RW) came into the stage of practical implementation.

The most large-scale remedial operations are to be performed at Facility No1 of the Federal state unitary enterprise "Northern enterprise for radioactive waste management" (FSUE "SevRAO") located on the Kola Peninsula in Andreeva bay.

The Federal Medical-Biological Agency of the Russian Federation (FMBA) is responsible for the regulation of radiation protection of workers and the public at operation of radiation hazardous facilities. Having in mind provisions of the regulative documents, FMBA of Russia presses for the consistent optimization principle introduction of the personnel radiation protection into practice of SevRAO facilities. The paper presents the findings of investigations performed by specialists from the State research centre – Institute of Biophysics regarding the assessment of the radiation monitoring system existing at SevRAO facility No 1; it also demonstrates some revealed features of the radiation situation generation on-site and in workshops at STS in Andreeva Bay; in addition, it proposes some recommendations relating to ALARA technique implementation in the course of SNF and RW management operations.

This work has been undertaken under NRPA financing.

Keywords Nuclear submarines, radioactive waste (RW), sites of temporary storage (STS), spent nuclear fuel (SNF), ionizing radiation source (IRS), OSPORB-99, NRB-99, ALARA principle

[1] State research centre – Institute of Biophysics (IBPh), Moscow, Russia

[2] Norwegian Radiation Protection Authority (NRPA), Norway

[3] Department for SNF and RAW management and decommissioning of nuclear and radiation hazardous facilities, Russia, Moscow

M.K. Sneve, M.F. Kiselev (eds.) *Challenges in Radiation Protection and Nuclear Safety Regulation of the Nuclear Legacy*,

1 Introduction

In «Justification of investments for the construction of the SNF and RW management infrastructure» [1] the following major stages of STS in Andreeva Bay decommissioning are being set:

- Design and construction of the required infrastructure for safe management of SNF and RW
- SNF and high level RW removal from the STS area
- Treatment and long-term storage of low level RW and
- Final remediation of the STS area

When constructing infrastructure for safe SNF and RW management, the following combines shall have to be constructed:

- SNF management combine, consisting of Building 153 – the shelter of tank of blocks dry storage (BDS), storage site for transport containers with SNF, auxiliary buildings, berth and other constructions
- RW management combine, consisting of the workshop for management of liquid radioactive waste (LRW), the workshop for management of solid radioactive waste (SRW), RW repositories and other constructions
- A set of auxiliary buildings, structures (canteen, garages, store-houses etc.) and engineered systems and communications (water supply, sewage with treatment plants, power supply, roads and sidewalks etc.)

At the stage of SNF removal, conditioned and damaged irradiated fuel assemblies (IFA) will be withdrawal and conveyed to PA "Mayak" for treatment. This withdrawal from BDS cells can be implemented according to the routine procedure. Old packages with non-discharged damaged IFA are conveyed to the special facility equipped with shielded chambers for IFA re-loading into protective sticks (pencil-box type) and for further treatment.

At the stage of RW treatment, about 26,000 m^3 SRW and 37,200 m^3 LRW are to be processed over the established period (15 years) [1]. Conditioned SRW and LRW (in packages suitable for following transportation and final isolation) will be allocated at the facility for temporary storage. Planned duration of conditioned RW storage on-site STS is 50 years.

Each stage of STS decommissioning has its inherent features affecting radiation situation and defining special conditions of work, therefore, there are some features of implementation of measures directed to the personnel protection optimization.

2 Abnormal Conditions of STS Operation in Andreeva Bay

In the course of long-term operation, the containment barriers of SNF and RW storage facilities deteriorated and lost partially their effectiveness, so that radionuclides were able to migrate into the environment, while workshops and the site of the facility No 1 became contaminated with radioactive substances above acceptable values. This and some other reasons led to abnormal conditions of STS operation now. These conditions, to a greater or lesser extent, will specify radiation situation at the early stages of STS decommissioning. Specific conditions inherent to Facility No 1, are mainly defined by [2]:

- Insufficient information regarding radiation and physical conditions of SNF and high level RW.
- Forced SNF and RW removal into BDS – buildings designed with other purposes.
- Insufficient information volume regarding damaged assemblies with SNF.
- Registration of increased levels of man-made radionuclide contents and external gamma radiation on-site STS, in the industrial buildings and constructions.
- Unique nature of technologies and equipment designed for SNF and RW management. Partial application is assumed of irregular equipment and instruments, which have no analogs within other atomic industry facilities.
- Necessity of practically simultaneous implementation of operations on contaminated sites directed at decommissioning some buildings and structures or their reconstruction and construction of new industrial buildings for work implementation of SNF management and RW treatment.
- Necessity of application of special personal protective equipment for workers, implementation of some radiation-hazardous operations under unfavorable meteorological conditions outdoors.
- A lack of sufficient staff of qualified personnel.
- Necessity of application of special equipment for collective protection of the personnel at SNF management.

With the purpose of the existing radiation situation assessment and actions development directed at guaranteeing radiation protection of workers and the public, specialists from Institute of Biophysics analyzed the pre-project documentation with respect to technology of SNF and RW management; they also carried out experimental examinations of the radiation situation parameters.

3 Radiation Situation in Areas for SNF Management (in BDS Locations for SNF Dry Storage)

The following conclusions had been resulted from the performed investigations:

- The equivalent dose rate (EDR) of external *gamma radiation* decreases with height when the distance from the radiation source (which is SNF located in cells buried within concrete) increases. This tendency is typical for all BDSs. Maximum EDR values are observed in BDS 3A location, where the EDR reaches 1.9 mSv/h at foot-level and 1.4 mSv/h at chest height.
- Gamma field in BDS rooms is of pronounced heterogeneous nature, because of the distribution of the main radiation sources (SNF cells) location, patchiness of contamination and the variation in height with distance from the source.

- Neutron dose rate can vary in a wide range mainly due to variation of contribution of thermal neutrons.
- Neutron dose rate values illustrate an absence of any significant contribution of this radiation type into dose of the personnel exposure within BDS rooms under conditions of SNF storage. This is not expected to change significantly during SNF removal operations.
- Work duration acceptable for the personnel in the BDS workshops, calculated in terms of the conservative approach (operation during a whole shift without organization of any protective actions) is limited. The strictest restriction of acceptable work duration is required for the personnel if BDS 3A, while the weakest one is required for workers in BDS 2A.
- As effective dose is the limiting factor defining acceptable work duration of the personnel in the workshops of each BDS, therefore some set of protective measures should be envisaged in order to guarantee radiation protection for workers at the stage of planning and organization of such kind of operations.
- Contamination activity by gamma-emitting nuclides is dominated mainly by Cs-137. The most contamination is observed on internal surfaces of cells, plugs (up to 2,000 Bq/cm^2). The least contamination levels are observed on surfaces of protective containers and on the floor, but the spread of values on these surfaces is wider: from few Bq/cm^2 to 620 Bq/cm^2. Here, the composition ratio of Cs-137 and Co-60 also varies in a significant range – from 3 to 2,170.
- The sample analysis suggests that at implementation of operations accompanied with intensive dusting acceptable activity concentrations of Cs-137 could exceed requiring the application of personal protective equipment for respiratory organs, dust control performance, surface decontamination or transformation unfixed radioactive contamination into foxed form with long-term fixation, as well as implementation of individual monitoring of internal exposure.

Under the existing abnormal conditions of SNF and RW management implementation, optimization procedures for assurance of occupational safety during operations (ALARA principle) are very important.

4 Assessment of Radiation Monitoring System: State and Perspectives

Optimization of the occupational radiological protection is based on reliable and valid information about radiation situation parameters and about individual exposure levels. These two issues are the functions of radiation monitoring.

5 Arrangement of Radiation Monitoring System

Radiation safety division (RSD) is responsible for radiation protection of workers at SevRAO facility No 1.

One of the main duties of the division is arrangement of periodic industrial (radiation) monitoring in process buildings and on-site STS, within HPZ and in SA. The nomenclature, extent and frequency of RM are defined in terms of:

- The industrial process characterization
- Presence of harmful industrial factors and
- Extent of their impact of human's health and habitat

and they are specified in the appropriate programs developed at the facility. Table 1 presents a listing of subjects and controlled indexes, monitoring of which are imposed as a duty on the radiation safety division.

Analysis performed by the specialists from SRC-Institute of Biophysics demonstrates that available radiation monitoring system promotes obtaining of full-scale information about radiation situation conditions at the industrial sites of SevRAO facility No 1 and complies with OSPORB-99 and NRB-99 requirements.

When going on to full-scale construction works and commissioning of the Combines for SNF&RW management, amount of radiation hazardous operations increases. In order to implement an optimization principle, under these conditions, the radiation monitoring system is reasonable to be enhanced, focusing on increasing monitoring extent and full-scale introduction of ASKRO system – depending

Table 1 Listing of the monitored subjects and controlled indexes

No	Name of monitored subjects	Controlled index
1.	Workshops, rooms, buildings and constructions located on the industrial site	Gamma dose rate
		β-fluence density
		Superficial contamination with α- and β-active substances
		Neutron dose rate
		Activity concentration and nuclide composition of radioactive aerosols
2.	Industrial site, routes of workers	Gamma dose rate
		β-fluence density
		Superficial contamination with α- and β-active substances
3.	Transport, package:	Gamma dose rate
	special transport with radioactive consignment; transport with general engineering consignment	Superficial contamination with α- and β-active substances
		Neutron dose rate
4.	Wastes:	Gamma dose rate
	RW	Contamination with α- and β-active substances
	general industrial waste	Neutron dose rate
		Specific activity and nuclide composition of LRW
5.	Equipment and materials	Gamma dose rate
		Superficial contamination with α- and β-active substances
6.	Personal protective equipment (external and internal surfaces)	Contamination with α- and β-active substances

upon established engineering procedure of work implementation. The radiation safety division will need additional organizational structure for verification and maintenance of dosimetry and radiometry equipment.

6 Structure of Personal Dose Monitoring System

Dose monitoring of external exposure to workers at SevRAO facility No 1 is subdivided into two types:

- Personal dose monitoring
- Group dose monitoring

Personal dose monitoring consists of occupational dose monitoring using personal dosimeter, which is attached to the particular person, or which this person has gotten for the period of either his/her management of ionizing radiation source (IRS) or his/her working under IRS exposure.

Group dose monitoring is implemented using group monitoring dosimeters as well as by means of dose calculation for the personnel working in the certain area (workshop).

Monitoring using personal dosimeters is obligatory either for workers, involved in direct operations with man-made IRS, or for those, whose conditions of work impose their staying in the sphere of IRS impact.

Within the facility, external dose monitoring using personal dosimeters is obligatory for the personnel "A" and "B" groups when they work within the controlled access area.

Individual external neutron doses during SNF management are calculated as a product of neutron dose rate over the particular period.

Beta-induced equivalent dose to the skin is calculated.

Radionuclide intake monitoring is performed using WBC spectrometry kit by means of direct measurements of workers. Monitoring is performed under standard geometries – «whole body» – determination of ^{137}Cs intake, «Lung» – determination of ^{137}Cs and ^{60}Co intakes, and «Thyroid» – determination of ^{131}I intake.

Monitoring frequency:

- The personnel "A" and "B" groups – annually, at the end of each calendar year
- The personnel involved into radiation hazardous operations, are subjected to monitoring before and after termination of work

Analysis performed by the specialists from SRC-Institute of Biophysics demonstrates that the existing system of personal dose monitoring (PDM) promotes obtaining full-scale information about occupational doses and complies with OSPORB-99 and NRB-99.

When full-scale construction and commissioning of Combines for SNF&RW management will start,

PDM system is reasonable to be improved, focusing on:

- Enhancement of personal dosimetry method with respect to external beta exposure to the skin; introduction of emergency neutron dosimetry
- Turn to application of thermo-luminescent dosimeters to assess external gamma doses
- Introduction of radionuclide intake assessment method in terms of the radionuclide activity concentrations in air of working area

7 ALARA Principle Application for Protection of Workers During SNF and RW Management

Under routine radiation source operation, Radiation safety standards (NRB-99) and OSPORB-99 require to follow the main radiation safety principles:

- Non-exceeding of authorized individual dose limits due to all radiation sources (dose limit application principle)
- Forbidden of all practices relating to radiation source application, if the resulting benefit both for individuals and for the society does not excess risk of possible harm due to additional exposure (justification principle)
- When using any radiation source, keeping individual dose and amount of persons exposed as low as reasonable achieved taking economic and societal factors into account (optimization principle)

Optimization principle is also called ALARA (**A**s **L**ow **A**s **R**easonable **A**chievable) principle. ALARA is the conception of dose limitation based on principles of minimization exposure levels taking economic and social reasonability into account.

ALARA principle is a constituent of the general safety culture of the facility, and its objective is minimization of risks. To establish the safety culture at the level of the facility, the following actions must be taken:

- To share the responsibility for safety and protection assurance
- To control work implementation
- To confirm qualification and to educate the personnel
- To arrange a system of encouragement and penalties
- To perform revisions, to make analytical reviews and comparisons

An optimization principle must be applied at all stages both of the manufacturing process arrangement and of the radiation facility operation; it must be implemented by means of:

- Generation of conditions for opening and implementation of each worker potentials (knowledge, skills, experience)
- Justified selection and preliminary planning of actions, implementation of which improves safety
- Preparedness for work implementation
- Analysis and evaluation of operations performed, account of experience gained

To manage development and taking actions for optimization principle implementation, ALARA group must be arranged at SevRAO facility No 1 (based on the current radiation safety division) under the chief engineer presidency.

ALARA group must deal with arrangemental issues of radiation hazardous operation implementation, including:

- Ways of the personnel involvement into work planning
- Preparedness for work under radiation hazardous conditions
- Control of protective measures
- Analysis and evaluation of results obtained, account of the experience gained

According to the Guidance R 2.6.1 29 – 07 [3], radiation hazardous operations include such SNF&RW management, at which predicted maximum individual effective dose to a worker, being calculated in terms of conservative assessments (including those which had been calculated without any protective/preventive measure account), can exceed 20 mSv per year.

When planning radiation hazardous operations, different options of their performance shall be considered. Options accompanied with the least dose costs are top priority. At that, the variant of operation implementation is preferable, which (taking account economic factors) ensures:

- The least individual occupational doses
- Minimum discharges and effluents of radioactive substances
- Minimum amount of radioactive waste generated

The conditions must be arranged, when the personnel deliberately select such ways, methods and work organization, which promotes achievement of the highest results (by quality and safety) under minimum time costs for work implementation.

The personnel, on its own, must take measures and methods of protection against ionizing radiations, such as:

- Protection by distance
- Protection by time
- Correct application of all type PPE
- Using computerized and automated apparatus, facilities and equipment

ALARA group in cooperation with the administration must develop a system of stimulation of work implementation under dose cost minimization accompanied with exactingness and compulsion (presence of the work leaders at radiation hazardous operation performance, periodic and unscheduled inspections, control implemented by the radiation safety division).

8 Conclusions

Specific abnormal conditions, existing on-site and in workshops of SevRAO facility No 1 (spent nuclear fuel and radioactive waste STS in Andreeva bay), call for increased attention to solving problems of radiation safety assurance fro the personnel.

Within the program of work implemented under NRPA financing, regulative base is building for optimization of radiation safety of workers during SNF management and RW treatment taking account special features of SevRAO real conditions and having in mind relevant international standards and regulations. At the early stage of the program implementation, the following regulative documents had been developed:

- The Guidelines «Personal dose monitoring of occupational exposure at SevRAO facility No 1»
- The Guidelines «Special features of ALARA principle application during SNF&RW management at SevRAO facility No 1»
- The Guidelines for monitoring «Procedure of radiation monitoring at SevRAO facility No 1»

Implementation of requirements from these documents is aimed at building of adequate regulatory basis for optimization of the personnel radiation protection using ALARA principle during SNF removal and RW processing at SevRAO facility No. 1.

In the nearest future (2008–2009), with the purpose to improve radiation monitoring and to plan radiation hazardous operations, two issues are reasonable to be elaborated. The first – arrangement of interactive mapping database on radiation situation on-site and in STS workshops. The second – arrangement of database on individual occupational external and internal doses. Implementation of these two projects will promote:

- To increase efficiency and quality of operator activity control implemented by the regulatory bodies
- To visualize results of many-year observation of radiation situation parameters (including gamma dose rate, site contamination with some radionuclides, beta fluence density, air contamination by radioactive aerosols etc.)
- To make prediction of individual occupational doses with the purpose of the personnel protection optimization during planning SNF&RW management operations using ALARA strategy
- As necessary, to carry out retrospective restoration of individual occupational doses
- In a short period, to select workers to be involved in the particular radiation hazardous operations and in consequence mitigation operations taking account accumulated and predicted doses

Acknowledgement The participating organizations would like to express their thanks and appreciation for the support of the Norwegian Radiation Protection Authority. The work described in this paper is funded by the Norwegian Government, through a Plan of Action promoting improvements in radiation protection and nuclear safety in Northwest Russia, implemented by the Ministry of Foreign Affairs (MoFA) and the NRPA.

References

1. Justification of investments into arrangement of infrastructure for SNF and RW management at STS in Andreeva bay (OBIN). FSUE «SevRAO» - VNIPIET, 2005.

2. Problems of radiation safety assurance under abnormal conditions. AV Simakov, OA Kochetkov, YuV Abramov, MK Sneve ea. In proceedings of the international conference «Up-to-date Challenges of the public radiation safety» 2006, Russia, Sanct-Petersburg, pp. 71–73.
3. Guidance «Hygienic requirements for radiation safety of workers and the public during planning and arrangement of SNF and RW management at SevRAO facility No 1» (R 2.6.1. 29 – 07)

List of Abbreviations

BDS	block of dry storage
EDR	equivalent dose rate
FSUE	Federal state unitary enterprise
IFA	irradiated fuel assembly
IRS	ionizing radiation source
LRW	liquid radioactive waste
NRB	Russian abbreviation of radiation safety standards
NS	nuclear submarines
PDM	personal dose monitoring
RM	radiation monitoring
RSD	radiation safety division
RW	radioactive waste
SNF	spent nuclear fuel
STS	site of temporary storage

Regulatory Control of Radioactive Waste in Sweden

H. Zika

Abstract This paper describes the legal base for radioactive waste management in Sweden, the regulatory authorities responsible for supervision and how they both relate to decommissioning operations in Sweden.

Keywords Swedish Radiation Protection Authority (SSI), Swedish Nuclear Power Inspectorate (SKI)

1 The Legal Base

- The Radiation Protection Act
- The Act on Nuclear Activities
- Act on Financing
- The Environmental Code
- Ordinances on Radiation Protection and on Nuclear Activities issued by the Government
- Regulations issued by the Competent Authorities

The Swedish Legislation is influenced by inter alia International Conventions (IAEA). The Joint Convention on the Safety of Spent Fuel Management and on the Safety of Radioactive Waste Management and the Convention on Physical Protection of Nuclear Materials are implemented in the Swedish Legislation. EU Legislation has either direct impact (EU Regulations) on Swedish Legislation or is to be implemented (Directives). Radiation Protection is implemented through the EU Basic Safety Standards (BSS). They are linked to the EURATOM Treaty.

2 The Swedish Competent Authorities

- The Swedish Radiation Protection Authority (SSI)
- The Swedish Nuclear Power Inspectorate (SKI)

Swedish Radiation Protection Authority (SSI), Sweden

M.K. Sneve, M.F. Kiselev (eds.) *Challenges in Radiation Protection and Nuclear Safety Regulation of the Nuclear Legacy,*

Both authorities are involved in the regulatory control of RadWaste and they will merge to a new authority, planned for start up 1 July 2008. The name of the new Authority is yet not decided. The Authority will consist of approximately 250 staff members, and it's mandate is expected to be similar to that for it's predecessors, inter alia supervision and licensing of nuclear activities such as RadWaste handling, nuclear safety, decommissioning. The financing system for RadWaste etc. is another responsibility.

3 The Swedish Waste Handling System

There is a concept that includes financing to deal with the decommissioning of nuclear fuel cycle facilities and to handle spent fuel and RadWaste, R&D etc. (Fig. 1)

Sweden has 13 NPP: s (three are under decommissioning), one Research facility with two MTR reactors (under decommissioning), a fuel fabrication plant and other installations. The funding system is a fundamental cornerstone in RadWaste handling. The funds are currently in the order of US$7 billion (7E9 $), and are financed through consumer fees on electricity. The new Authority will be responsible for this funding system, including the build-up and use of the funds. SKB is a company owned by the power utilities. One of it's tasks is on behalf of its owners to provide a political acceptable system for final repositories for spent fuel and nuclear waste.

Fig. 1 Nuclear facilities in Sweden

Today there are licensed repositories for very low level waste (VLLW) and low and intermediate level short lived waste (LILW-SL). An application for a final repository in deep bed rock for spent fuel is expected to be filed by SKB in 2009. Further repositories for decommissioning waste from NPP: s and long lived waste (LILW-LL) are under planning.

4 Regulatory Control and Supervision of RadWaste

SSI, SKI and effective from next year the new Authority are responsible for both licensing and supervision of all aspects of RadWaste handling (Figs. 2 and 3). That includes the production of the waste and handling, transporting, storing etc. of the material. This is done both by sending inspectors onsite, and through document controls. Of special interest is the quality aspect of safety reports and other documentations. Special inspections performed by the authorities are dedicated to the Quality Assurance (QA) Systems of the license holders.

Currently the authority SSI is undergoing a certification process according to ISO 9001, in addition to an earlier ISO 14001 certificate (Environmental Certificate). Both SSI and SKI have regular internal QA audits, but SSI will also be audited by a external licensed review companies (Fig. 4).

Fig. 2 Inspection of a RadWaste transport: a 320 t steam generator at Ringhals NPP

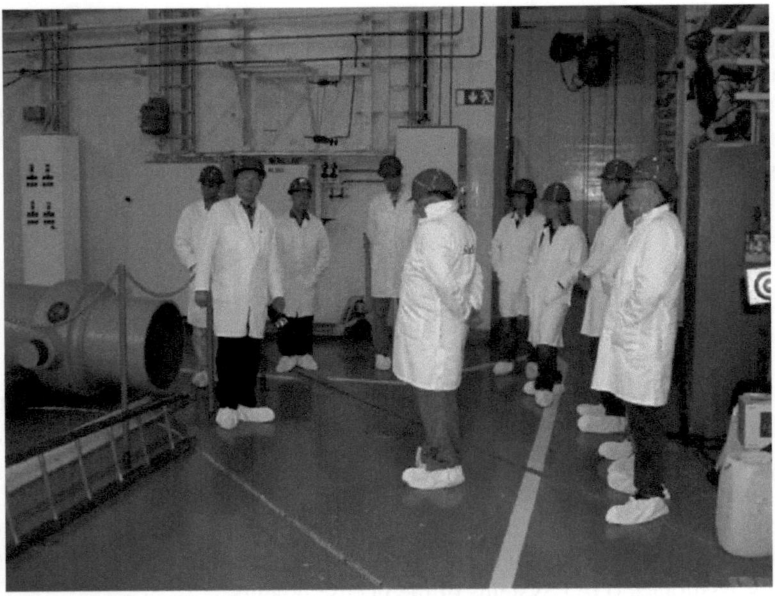

Fig. 3 SSI and SKI inspectors at a decommissioned PHWR at Ågesta

Fig. 4 Control measurement inside a decommissioned facility prior to free release

MIX
Papier aus verantwortungsvollen Quellen
Paper from responsible sources
FSC® C105338

If you have any concerns about our products,
you can contact us on
ProductSafety@springernature.com

In case Publisher is established outside the EU,
the EU authorized representative is:
Springer Nature Customer Service Center GmbH
Europaplatz 3, 69115 Heidelberg, Germany

Printed by Libri Plureos GmbH
in Hamburg, Germany